# INDIAN ECONOMY and the Information and Communications Technology (ICT)

# Indian Economy

## and the Information and Communications Technology (ICT)

*By*

Seema Kapoor

Assistant Professor of Economics,
S.A.V. Jain College, Ludhiana

New Century Publications
New Delhi, India

NEW CENTURY PUBLICATIONS
4800/24, Bharat Ram Road,
Ansari Road, Daryaganj,
New Delhi – 110 002 (India)

Tel.: 011-2324 7798, 4358 7398, 6539 6605
Fax: 011-4101 7798
E-mail: indiatax@vsnl.com • info@newcenturypublications.com
www.newcenturypublications.com

*Editorial office:*
LG–7, Aakarshan Bhawan,
4754-57/23, Ansari Road, Daryaganj,
New Delhi – 110 002

Tel.: 011-4356 0919



First Published: **2013**

ISBN: **978-81-7708-345-3**

Published by New Century Publications and printed at Salasar Imaging Systems, New Delhi.

Designs: Patch Creative Unit, New Delhi.

PRINTED IN INDIA

*DEDICATED TO MY FAMILY*

# About the Book

India has emerged in recent years as the most preferred destination for outsourcing of Information and Communications Technology (ICT) services. The vibrant ICT industry is contributing immensely by providing information about latest technology and international business practices. Various sectors of the Indian economy—industry, finance, trade, transport, health and education—have adopted information technology in a big way.

In fact, ICT industry has become India's premier growth engine, crossing significant milestones in terms of revenue growth, employment generation and value creation. The rapid growth of the industry has been attributed to availability of human capital, entrepreneurship, social networking, favourable policy environment and entry of multinationals.

This book explains and examines in detail the role and contribution of the ICT industry in various sectors of the economy of India. It provides an analytical framework to evaluate the competitive strength of ICT industry in India.

# Author's Profile

**Dr. Seema Kapoor** is presently Assistant Professor of Economics, Shree Atma Vallabh Jain College, Ludhiana. She did her M.A. (Economics) from D.A.V. College, Amritsar and obtained M.Phil. and Ph.D. degrees from Guru Nanak Dev University, Amritsar. With 17 years of teaching and research experience, she has published a number of papers in reputed journals including *World Journal of Social Science* and *Journal of Innovations.* She has attended and presented papers at various national and international conferences/seminars.

# Contents

# Preface

The Indian ICT industry has steadily moved up the value chain and is now fully capable of envisioning and developing a whole range of value added services that can be delivered through computers, mobile phones and other devices. The industry can create and sustain a service delivery platform for such services thereby extending the global reach of the entire service industry.

The advent of cloud computing technology has thrown up another wide range of possibilities for India. Indian companies have already started creating new products and services incorporating their own IPs and patents. The stage is now set for a quantum jump in this field. Another positive aspect is that there is a growing trend of entrepreneurship and product/service innovations. Information technology (IT) and telecommunications are the twin sectors that best epitomize what modern, resurgent and young India is capable of accomplishing. Achievements in the recent past in the IT and telecom sectors have been spectacular and unprecedented in India's history. Yet, it is the future that holds even more breathtaking possibilities. A fascinating opportunity has now arisen that rests on the foundations of what has been achieved so far in IT and telecom.

In the global economy of 21st century, a country's development policy challenges require it to use knowledge more effectively to raise the productivity of agriculture, industry and services and reduce poverty. Knowledge is more important today than ever, as the twin forces of globalization and technological advances are spurring an ongoing knowledge revolution. The term "knowledge economy" reflects this increased importance of knowledge for economic development. A knowledge economy is one that creates, disseminates, and uses knowledge to enhance its growth and competitiveness.

India has endeavoured to build and consolidate its

knowledge base since Independence in 1947. Indian planners envisaged the strategy to attain economic and technological self-reliance by investing perceptibly in education, R&D and health sectors. A gigantic network of research laboratories and institutions, coupled with a chain of engineering colleges and IITs has enabled India to forge a formidable reservoir of technically skilled workers. Globalization of the Indian economy since early 1990s has resulted in a massive home-grown technically professionalized manpower. With increase in the use of ICT in knowledge intensity sectors of the economy, India has a tremendous growth impetus to emerge as the fourth largest economy in the world.

Over the past decade, the Indian ICT sector has become the country's premier growth engine, crossing significant milestones in terms of revenue growth, employment generation and value creation, in addition to becoming the global brand ambassador for India. Indian IT service segment has evolved from application, development and maintenance, to emerge as full service player providing testing services, infrastructure services, consulting and system integration. IT outsourcing has exhibited a strong growth, in line with the global trend, due to increased spending in the remote infrastructure management, application management and testing. The rapid growth of the industry has been attributed to availability of human capital, entrepreneurship, social networking, favourable policy environment and entry of multinationals.

This book explains and examines in detail the role and contribution of the ICT industry in various sectors of the economy of India. It provides an analytical framework to evaluate the competitive strength of ICT industry in India.

I place on record my deep sense of gratitude to Dr. Vikram Chadha, Professor and Head, Punjab School of Economics, Guru Nanak Dev University, Amritsar for his invaluable guidance and encouragement at various stages of the completion of this work. I extend my thanks to Shri Desraj Kondal (IRS), Joint Commissioner of Income Tax, Amritsar;

Shri Rajinder Kumar, Deputy Director, Department of Commerce, New Delhi and Dr. Roukmani Gounder, Associate Professor of Economics, Messey University, New Zealand for their valuable suggestions and help.

I sincerely acknowledge the affection and encouragement showered on me by my mother. Heartfelt gratitude is also due to my husband Shri Ravi Kapoor and loving son Varun Kapoor who have always been supportive of my academic pursuits.

Finally, I thank my publisher New Century Publications, New Delhi for unfailing courtesy and enthusiastic co-operation in the production of the book.

***Ludhiana***
**November 2012**

**Seema Kapoor**

# 1

# WTO, Globalization and Information and Communications Technology (ICT)

Over the past fifty years, the role of trade in the world economy has increased dramatically. Trade has grown much faster than output and most of countries that have achieved fastest economic growth have done so rapidly by increasing their participation in the world trade. Import substitution policies—widely used in the 1950s, 1960s and 1970s were much less successful than the export-oriented policies used in East Asia's high-growth economies. As a result, policy makers in many developing countries started adopting policies for more open trade during the 1980s. By the late 1980s, nearly all centrally planned economies had made reforms in foreign trade and investment for their development programs.

These reforms had led to significant changes in developing countries' involvement in international trade. Until the mid-1980s, most developing countries relied primarily on commodity exports, which exposed them to high volatility and decline in commodity prices and more dependence on imported manufactures. However, since early 1990s, due to the increasing global integration and multilateral trade negotiations many developing economies increased their manufactured exports. By the late 1990s, about 80 percent of exports from developing countries were manufactured goods. Thus, developing countries are now active in global markets for manufactures and services.

## 1.1 World Trade Organization (WTO) and Globalization

The General Agreement on Tariffs and Trade (GATT) was established in 1948 in Geneva to pursue the objective of free trade in order to encourage growth and development of all member countries. The principal purpose of GATT was to ensure competition in commodity trade through the removal or reduction of trade barriers. The first seven rounds of negotiations conducted under GATT were aimed at stimulating international trade through reduction in tariff barriers and by reduction in non-tariff restrictions on imports imposed by member countries. The 8th round of Multilateral Trade Negotiations, popularly known as Uruguay Round was started in September, 1986 at a special session of GATT Contracting Parties held at Ministerial level.

World trade had undergone a structural change during the five decades since the establishment of GATT in 1948. In the case of trade, liberalization was pursued multilaterally through successive GATT negotiations.

Increasingly, bilateral and regional trade agreements became an important aspect of (preferential) trade liberalization as well. World Bank and the GATT played an important role in the process of globalization. They have provided cohesion and greater coherence to international economic policymaking.

Thus, Uruguay Round (UR) contained the mandate to have negotiations in 15 areas. In Part I, negotiations on Trade in Goods were to be conducted in 14 areas and in part II negotiations on Trade in services were to be carried out.

The process of liberalization culminated into WTO, which incorporated broader parameters of policy and envisaged a liberalization rule-based trade regime. The establishment of Word Trade Organization (WTO) in 1995 as a watchdog on Trade marked an important turning point and the beginning of the new era in the world economic management. As on August 24, 2012, 157 countries are members of WTO.

WTO contains various trade procedures acceptable to both developed and developing countries. It is based on the principles of non-discrimination and provides predictable basis of trade. In the globalized context, WTO also takes note of matters incidental and ancillary to trade and investment such as, employment standards, reduction of tariff rates, investment protection, trade and environment, and trade and employment.

## 1.2 Globalization of World Economies

The rapid growth in openness of world economies has been alternately described as globalization/internationalization. While there is no universally agreed definition of globalization, economists typically use the term to refer to international integration in commodity, capital and labour markets. Using integration in these markets as the benchmark, it is clear that globalization is not a new phenomenon. Since the mid-19th century, there have been at least two waves of globalization (WTO, 2008). The first phase began around the mid-19th century and ended with the onset of global depression. The second scenario began in the aftermath of World War II (WWII) and continues until today. In both these episodes of globalization, rapid trade and output growth went together with major shifts in the relative size of the economies involved.

International trade after WW II entered a long period of record expansion with world merchandise exports rising by more than 8 percent per annum in real terms over the 1950-73 periods. Trade growth slowed thereafter under the impact of two oil price shocks, a burst of inflation caused by monetary expansion and inadequate macroeconomic adjustment policies. In the 1990s, trade expanded again more rapidly, partly driven by innovations in the information technology (IT) sector. Despite the small contraction of trade caused by the dotcom crisis in 2001, the average expansion of world merchandise exports continued to be high.

Thus, globalization refers to doing business abroad and adapting to international business standards to conduct business transactions with others. Globalization is a policy or a system favouring or promoting globalization or the condition of being globalized. The interdependence of countries under globalization helps to promote cross border flow of goods and services, capital and technology. Enterprises with global mindset can divert their energies to the expansion of business opportunities beyond domestic markets; attract new customers; enhance their market share and maximize their profit margins. The economic liberalization of developing countries have complemented the globalization process by generating competition and comparative cost advantages among business firms. In market, companies learn to meet the needs of diversified cultural group, modify business strategies and gain management knowledge (Venkateswarlu, 2004). Globalization has become a reality and countries both developing and developed have no choice but to adjust to that reality.

Globalization or openness of world economies includes reduction in trade barriers, transport costs and communication costs. Related factors include the increasing importance of trade in manufactures and the fragmentation of production process which leads to much more international trade in components and services. Well-designed globalization can promote technology and development. In fact, it is only with a certain degree of openness that one can get to the ideas that are necessary for adoption of technology.

**1.2.1 Globalization and Growth:** Globalization has provided the backdrop for the strong economic growth in recent years. There are three major aspects of globalization, namely international trade, foreign direct investment and labour migration. It has been widely recognized that these three dimensions of globalization affect, to a greater or lesser extent, the level and distribution of wages. The share of wages in GDP has declined and the growing wage inequality seen in several regions around the world. The wage share declined over the period 1995-2007 in two thirds of the developing countries. The only exception was the Latin American region, where some countries witnessed an increasing wage share.

In addition to it, the distribution of wage among wage earners has worsened. The average wage of the top 10 percent of wage earners (in relation to the bottom 10 percent) is found to have increased in 70 percent of the countries (Hoeven, 2010). The underlying reasons for the increasing wage inequality vary across countries. Wage inequality is growing due to deterioration in the lowest wages and faster increase in top wage earnings than in other wage groups. The wage share has declined faster in countries with a higher openness to international trade. The wage gap between women and men is also still high and is closing very slowly. This is disappointing in the light of women's recent educational achievements and the progressive closing of the gender gap in work experience (ILO, 2008-09).

Another noticeable trend over the last two decades has been the internationalization of the production process. Today, there are some 82,000 multinational corporations (MNCs) with 810,000 affiliates in the world (Hoeven, 2010). These companies play a major role in the world economy. For instance, exports from foreign affiliates of MNCs are estimated to have grown from about one quarter of total world exports of goods and services in 1982. These MNCs are dominated by a smaller number of large firms. The largest 100 MNCs account for 11 percent of total employment by MNCs and for about 4 percent of world GDP. Over the last 15 years, the largest MNCs have undergone a rapid process of internationalization. There has also been an increase in the proportion of companies operating in the service sector and of firms based in developing countries.

Technological innovation has been one of the main forces driving global integration. Massive investments in road infrastructure and revolution in information and communication technology have allowed large growth of trade in manufacturing and commercial services since 1980. New products such as the microprocessor, the personal computer, the cellular phone, internet and World Wide Web have contributed to profound socio-political and economic transformation.

World exports of commercial services rose at an annual

average rate of 11 percent during 2000-06 from a lower annual rate of 5 percent during the second half of the 1990s. The fastest growths of the major services categories include transport followed by travel (10 percent) and other commercial services (10 percent). The share of other commercial services in the total value of exports was 51 percent, while travel and transport each represented 25 percent and 23 percent respectively in 2008 (World Bank, 2008).

## 1.3 Role of ICT in the Globalized World

In the global economy of 21st century, a country's development-policy challenges require it to use knowledge more effectively to raise the productivity of agriculture, industry and services and reduce poverty. Knowledge is more important today than ever, as the twin forces of globalization and technological advances are spurring an ongoing knowledge revolution.

This dynamic process is creating constant restructuring at the global, country, sector, and firm levels. It has raised tremendous possibilities for enhancing growth and competitiveness, but it also carry the risk that countries or firms and organization will fall behind if they cannot keep up with the pace of rapid change. Consequently, countries' competitiveness depends more than ever on their ability to access, adapt, utilize, and create knowledge.

The term "knowledge economy" reflects this increased importance of knowledge for economic development. Use of knowledge has been always a critical ingredient of economic success; however, its importance has increased in recent times to the point that knowledge has become the key driver of economic competitiveness and success. Improved knowledge has led to increased productivity and the creation and application of new technologies has increased the range of products and services. It has brought revolutionary change to virtually all markets and sectors.

## 1.4 Knowledge and Knowledge Economy

Modern developed societies have become information based knowledge societies, where knowledge accumulation has taken place through a long history of discoveries, inventions, learning and improvements in the past. New ideas methods, skills and knowledge are intangible assets of a knowledge economy. They have not only contributed to physical capital accumulation and accelerated growth of these societies, but such immaterial investment in knowledge building through education, and research and development (R&D) has further accentuated knowledge accumulation at a still accelerated rate. So, knowledge manifested in new ideas, skills, methods, R&D, and learning-by-doing, have become invisible and subtle source of socio-economic growth of societies.

Knowledge can be described as a public good. It is seldom quantified or priced; it is sometime codified, but more frequently a tacit; and in any more case, it is difficult or impossible to observe. All measures of knowledge are indirect either inputs to (years of schooling, manuals) or outputs of (human capital, patents) its accumulation (Navretti and Tarr, 2000). It is non-rivalrous and non-excludable in many respects and there are large externalities associated with knowledge.

Pure public good is a good that is non-rivalrous and cannot be excluded from certain users. Ideally, knowledge is to some extent excludible so it might be considered an impure public good. Though efficiency in practice requires that there will no be charge; but with no charge there will not be any incentive for the firms to produce knowledge. Therefore, to provide knowledge privately, there must be some form of protection. Hence, though knowledge is not pure public good, but certainly there are externalities associated with it .The knowledge, by definition, gives rise to a form of increasing returns to scale that may undermine competition. These concerns are reinforced by the large network externalities, such as those associated with the use of computer languages

(Stiglitz, 2003).

Thus, by knowledge we mean the accumulated store of information and data we carry around our minds, together with our own interpretation of them based on reasons, instincts and prior experience. It can include 'know what' (knowledge about facts), 'know why' (scientific knowledge of the principles and laws of nature), 'know how' (skill or capacity to do something) and 'know who' (information about who knows what and how to do what).

Knowledge has to be considered at the core of technological development. Several forms of knowledge, including technical knowledge, knowledge of quality, knowledge pertaining to context, knowing how to learn and knowing how to marshal knowledge, would be required to be integrated for initiating and maintaining development.

In the history of civilization, four basic types of economies have been identified. The primitive economy, in which hunting was the only source of livelihood, was followed by the agricultural economy. The transition from primitive economy to agricultural economy took a few thousand years. The industrial economy followed the agricultural economy and transition took merely 200 years. Digital revolution, the internet has given birth to the knowledge economy. The most remarkable feature of the knowledge economy is the speed with which it has developed us. It took only 25 years for the knowledge economy to develop. But what is meant by knowledge economy and how can we define it?

Dr. Paul Romer, a Stanford economist developing the theory expounded by Professor Robert Solow, a Nobel laureate in the 1950s, put forward the New Growth Theory. According to him, technology and knowledge on which it is based is an intrinsic part of the economic system. Economic growth is driven by accumulation of knowledge. In other words, the knowledge economy is the one in which generation and exploitation of knowledge plays a prominent role in the creation of wealth (Bezborah, 2006).

The knowledge economy differs from other economies in a significant way in that service rather than agricultural or manufacture of goods dominates the source of employment. Consequently, the new commodity in the knowledge economy is skilled educated work force. Work force consist of not only people with formal education but also those who have acquired expertise through tacit knowledge which is knowledge gained by experience. The knowledge economy emphasizes tacit knowledge, which is as important as institutionalized knowledge.

Economic and social organization in a knowledge economy revolves around information. Since information is the raw material, information and communication technology play a crucial role in shaping the knowledge economy. These two technologies have the potential to increase wealth generation in the knowledge economy by adding value to existing products and services. Knowledge has become a central, productive and strategic asset, which can give way to sustainable development. The driving forces of this life age, which are playing the major role of changes, are:

1. An exponential increase of competitiveness at all levels because of globalization of the world economies.
2. Rapid changes in technology, especially in the internet and telecom technology.
3. Growing people interest in investment for science and technology researches.
4. The growing demand for refinement of skills.
5. The growing concern for protection of intellectual property rights.
6. Growing knowledge-intensive business.
7. A genuine increase in the commitment for a participative work-culture.
8. Rapid growth in the networking strategies.

Thus, by knowledge economy, we mean an economy, which revolves around creating, sharing and using knowledge and information to create wealth and improve quality of life.

On the same lines, Srivastava (2001) finds that in knowledge economy, intellectual capital, especially the power and product of mind, is the supreme source. Intellectual capital is made up of human and knowledge capital. Human capital comprises individual talents and knowledge acquired through education, training and experience. Knowledge capital is the documental knowledge available in such forms as reports, patents, software, book, papers, articles etc. The effective interaction of the two kinds of intellectual capital is essential for maximizing the productivity.

The knowledge economy is often taken to mean only high technologies industries or information and communication technologies (ICTs). It would be more appropriate, however, to use the concept more broadly to cover how any economy harnesses and uses productivity of agriculture, industry and services for overall welfare. Therefore, in such an economy great potential exist for increasing productivity by shifting labour from low productivity and subsistence activities in agriculture, informal industry and informal services activities, to more productive modern sector, as well as to new knowledge-based activities—and so doing, to reduce poverty and touch every key member of society.

In knowledge economy, knowledge is converted into wealth through the process of innovation. An efficient innovative system in the country involve creation of clusters, which are network of: (a) inter-dependent firms; (b) technology and knowledge producing institution like universities, colleges and institute, research institutes, and technology providing firms; (c) institution like think tanks, providers of technical or consultancy services; and (d) customers linking the value-added production chain.

The concept of cluster involves all form of knowledge sharing and exchange. Thus, an innovative system with its cluster, taps into the growing stock of global knowledge, assimilate and adopt it to local needs and finally create new knowledge and technology. In other words, knowledge economy leads to a developed economy.

## 1.5 Knowledge and Economic Growth

In the 21st century, the continuous acquisition and development of knowledge is not merely advantageous; it is indispensable. As knowledge accumulates and diffuses across the planet, it becomes more plentiful and less costly to acquire. As the knowledge component in every industry becomes larger and larger; comparative advantages in labour costs and natural resources available becomes progressively less important. To acquire and maintain a globally competitive knowledge base in its core business activity is a principal challenge to every modern business firm. More than any other factor of production, knowledge capital is becoming the most critical factor for nations' long-term competitive success.

Product differentiation, quality improvement, modernization, investment, environmental friendliness, innovation in materials, structure and design, increasing value addition and product durability are all vital aspects of dynamic and competitive industries. All are dependent upon the acquisition of continuous growth in knowledge. The application of knowledge is not limited to production, however. Any viable industry today is using highly sophisticated knowledge to revolutionize its marketing channels, to improve human resource development, to increase the efficiency of management, to strengthen corporate accountability and governance, to lower capital costs, to reduce all kinds of risks and challenges, to build a supportive policy environment, to enhance brand identity, consumer trust and customer loyalty.

The requirement for such activities increases as the kinds of knowledge demanded become more specialized and complex, as the pace of technological innovation accelerates, and as competitive advantages becomes ever more demanding.

Countries with higher per capita income hold larger stocks of accumulated knowledge. However, most advances in knowledge are not attained at once and then slowly spread, rather, today's' state-of-the-art technology is the outcome of many building blocks, and it embodies many ideas. To assemble many ideas, one generally needs to involve many people. If so, the growth of knowledge is essentially an interactive process, and hence one that is essentially inseparable from its diffusion. Therefore, diffusion and growth of human capital is a joint product (David and Loewy, 2002).

It has been found that it is the combination of different ideas that give rise to still better idea. As such, the acquisition of knowledge is a search process. Nevertheless, knowledge is unevenly distributed in any economy. If each agent knew exactly the same things, no amount of communication among agents would increase the amount of knowledge per person. Evidently, then, international spillovers are the result of heterogeneity. Thus, heterogeneity plays an important role in the diffusion and growth of knowledge.

The link between growth and accumulation of knowledge has been tested by few rigorous empirical studies. These empirical studies are based on knowledge variables as endogenous to growth in open economies. This literature recognizes two main mechanisms of knowledge accumulation. The first is that trade may change a country's pattern of specialization. Learning is faster if the country specializes in goods with higher learning potential (Krugman 1987; Stokey 1991). The second mechanism is that trade in goods and factors of production may open sources of technological inputs (Grossman and Helpman 1991; Rivera-Batiz and Romer 1991a).

Education is fundamental enabler of knowledge economy. Higher levels of education promote economic growth and development. Thus, investment in education is a policy priority in many developing countries. Education is charged with the function of generating basic skills across the population, and advanced cognitive skills amongst a portion of population. Education is both the seed and the flower of economic development. The empirical macro growth literature typically specifies growth as a function of the initial level of education. The effect of initial level of education on growth has been widely interpreted as an indication of large externalities from the stock of a nation's human capital on growth. In the words of Barro (2001), "On impact, an extra year of male upper level schooling is estimated to raise the growth rate by a substantial 1.2 percent points per year".

He found that workers with educational background would be complementary with new technologies; the results suggest an important role for diffusion of technology. The accumulation of efforts devoted to schooling and training lead to development of human capital. Human capital plays an important role in growth process. The skills that an individual acquire may be applied to an increasing set of production technology, in which value of human capital will continuously rise through time.

The development and integration of new knowledge leads to successful innovation. A process of continuous innovation and upgrading of technology acts as a key to dynamic growth in any economy in modern times. The higher speed of technology innovation will enable the developing country to have a higher return to capital accumulation and industrial upgrading. It will reallocate labour and other resources from low value-added industries to higher value-added industries. Innovations enable firms to manufacture state-of-the-art products and thereby move up the quality ladder relative to competitors. It will slow down import penetration and allow individual nations to grow more rapidly (Yifulin, 2004).

## 1.6 Interface among Knowledge, Globalization and Growth

Since early 1980s, and more specifically since early 1990s, the whole world is integrating as a global market economy, characterized by stimulated and unrestrained trade, capital and technology flows across geographical frontiers of countries. Globalization has induced the economies to exchange based on comparative advantage. For improving their competitiveness, economies are trying to reinforce their knowledge base and thus invest heavily in education, health and R&D. With accumulated knowledge, modern economies are able to realize high growth projections. There are two major interpretations of the term globalization. The two appear to conflict with, and even oppose, one another.

The first, and the mainstream view, takes globalization as a rational outcome of global economic "progress". This view describes it as a process of: (i) expansion of free trade; (ii) more integration of all economies of the world; (iii) more communication amongst different regions; (iv) faster diffusion of knowledge; (v) increasing mobility of labour and capital; (vi) increasing competition; and (viii) emergence of global economy-global market.

A second, and dissenting view, lays more emphasis to the power and ownership aspects of the globalization process. It looks at the process as being increasingly dominated by multinational corporations (MNC) and global institutions like the World Bank, International Monetary Fund (IMF), and World Trade Organization (WTO). Thus, according to this view, the present globalization process is in fact the manifestation of power by large corporate entities, which are mainly cantered in the North. They also argue that the worldwide expansion of the capitalist drive for profit has endangered the environment, people's lives, local authority, bio-diversity as well as diversity of culture. They also insist that an alternative to the present globalization process exist, which is, a real and peoples' globalization.

Nevertheless, it would be more plausible to argue that the globalization process, in fact, embraces elements of both views. It is essentially capitalism, which is advancing on a world-scale and is integrating peripheral economies into a single global system. Since the last decade, when globalization became a "buzzword", world trade has been expanding, and most economies have become more integrated into single global economic system. Information and communication system too have developed at a faster rate because of the rapid development of information technology. Now the opportunity for greater diffusion of knowledge has emerged, and capital has become more mobile. These facts support the mainstream version of globalization (Muhammad, 2006).

In other words, globalization is a complex process. Globalization not only leads to specialization and concentration of commodity production, it also requires large sophisticated service industries. In Britain, international services displaced potential manufacturing growth. The knowledge economy resides in service industries (Harley, 2000).

Well-designed globalization can promote technology and development. Trade links and FDI often bring access to markets and to new technology. With a certain degree of openness, one can get exposure to the ideas that are necessary to close the knowledge gaps. Globalization has made knowledge of world much more easily available than ever before. Thus, globalization can act as an important vehicle through which technology is promoted, and by promoting technology, development too can be stimulated.

The relationship between trade openness and economic growth has been subject of numerous empirical studies. Most uncover a positive empirical association between trade openness and per capita income growth, until recently, few economists challenged the findings. Some theories stress technological spillovers and the international transmission of knowledge as a source of growth for open economies. Higher levels of output are attained when countries specialize according to comparative advantage. So growth rate can be expected to increase in the transition that follows a liberalization path. The increased degree of market competition resulting from a wider scale of market interactions yield further gains in efficiency. More generally, trade openness increases size of market and allows economies to realize the benefits of increasing returns to scale. Yet another set of theories highlight that trade openness may create incentives for government to adopt less distortions in domestic policies and more disciplined type of macroeconomic management (Waczirag, 2001)

As open economies tend to specialize according to their comparative advantage, theory points to a lower degree of price distortions in open economies. Price distortions have been shown to adversely affect factor accumulations and growth. Factor accumulation may well also be of crucial importance. Trade liberalization may provide the type of big push effect on capital accumulation that was required for less developed countries to move from low growth equilibrium to a path of sustained industrialization. Another investment channel for growth under openness is the import of capital goods that were previously unavailable. Thus, it removes structural constraints on investment. Capital goods imports, which make up sizeable proportions of the imports of many recently liberalized developing economies, also embody more recent technologies which are a further source of growth.

Increased international trade can enhance the technical knowledge transmission to domestic economy. It will make easier for domestic producers to imitate foreign technologies and incorporate that knowledge in their own productive processes. This will enhance their capacity to compete with more advanced economies on world markets. Such a pattern was certainly part of the East Asian growth miracle, characterized by broad transformation in the product composition of output and exports from agriculture to heavy industry and finally to high-tech goods, through the imitation of technology originating in industrial countries.

A second channel for greater technology transmission is foreign direct investment (FDI). FDI often transmits advanced types of technology, either through capital goods imports that are later imitated or through the diffusion of knowledge and expertise. However, trade openness is associated with greater levels of FDI. FDI may act as a substitute for trade; because foreign investment is used to set up plants producing goods that cannot be imported because of trade restrictions. Foreign direct investment (FDI) may complement rather than substitute for trade openness. Indeed, evidence suggests that open economies attract more FDI than closed economies.

Exporting is another channel through which firms based in open economies can acquire foreign knowledge. There is abundant evidence that exporters are on average more productive than non-exporters (e.g., Tybout, 1992; Bernard and Jensen 1999). The analysis has shown that there is evidence for a strong learning from-exporting effect.

This underlines the importance of identifying the major determinants of successful technology diffusion. Among the many that have been proposed, two stand out: human capital and research and development (R&D) expenditures. Both are associated with the notion of absorptive capacity, a firm or country needs to have a certain type of skill in order to be able to successfully adopt foreign technology. These skills can come, first of all, in the form of human capital. They can also be in the form of R&D spending. It has been argued that R&D investments are necessary for a firm to acquire outside technology.

This is because R&D is critical to enabling the firm to understand and evaluate new technological trends and innovations. Eaton and Kortum (1996) find that inward technology diffusion is increasing in the level of a country's human capital. In Caselli and Coleman (2001) study, inward technology diffusion is captured by imports of office, computing, and accounting machinery, on the ground that many countries do not have a domestic computer industry, so that computer technology comes necessarily from abroad. They find that computer imports are positively correlated with human capital. There is evidence that the relative importance of international technology diffusion has been increasing with the level of economic integration in the world. While there is no consensus yet on the exact magnitude of spillover benefits, it is clear that well-functioning markets and an undistorted trade and foreign-investment regime are conducive to these learning effects (Keller, 2004).

### 1.7 IT Proliferation under Globalization

The above evidence is adequate to establish a direct relationship among globalization, growth of knowledge and economic growth. As globalization strengthens, a capacity is built up to generate more knowledge, and as knowledge accumulates through diffusion and spillover effects of unhindered trade and exchange, faster growth is induced to catapult economies on higher growth trajectory.

During the last decade, the world has become more integrated through trade and technology. Due to globalization, trade of technology and technology embodied capital, and consumer goods and services have witnessed impressive growth. This changing economic scenario has far reaching implications for employment, productivity, and efficiency. Today, we are living in the age of information and communication. The rapid growth of information and communication system is increasingly recognized as more important, in terms of major technological revolutions of the past two centuries. IT is permeating all aspects of business and society.

A consensus has emerged that faster growth can be traced at least in part to the effects of IT revolution. These effects can occur through three transmission channels. First, rapid technological progress in the production of IT goods raises total factor productivity growth in IT manufacturing industries. Second, falling prices of IT goods induce an IT investment boom. This adds to the capital available to workers, making them more productive. Third, as argued by David (1990), IT typically is general-purpose technology, which is pervasive and will spread with a time lag. It facilitates and induces firms to introduce more efficient organizational forms. It will eventually boost production across the economy. In other words, effect of IT on growth occurs through IT production, IT investment, and IT-induced innovation in services (Timmer and Ark, 2005).

The definition of IT investment, broadly, include three CT asset types, namely, 'computers', which comprises the whole category of office, accounting, and computer equipment; 'communication equipment' which include radio, TV, and telecommunication equipment; 'software' including pre-packaged own account and customized software.

Changes in information technology costs and capabilities lead to a cluster of changes in work organization and firm strategy that increases demand for skilled labour. It has been found that information technology capital is disproportionately associated with intangible assets like the cost of developing new software, populating a database, implementing a new business process, acquiring a more highly skilled staff, or undergoing a major organizational transformation, all of which go unaccounted on a firm's balance sheet.

Such assets are large, potentially several multiples of the measured information technology investment. They are unmeasured in the sense that they do not appear as a capital asset or as other component of firm input, although they do appear to be unique characteristic of a particular firm. They have more effects in long run than short term, suggesting that multiple years of adoption and investment is required before their impact is maximized (Bressnahan et al., 2002).

Another important finding is that information technology investment is greater in organizations that are decentralized and have a greater investment in human capital. A variety of industry level studies also show a strong connection between investment in high technology equipment and demand for skilled and educated workers (Berndt et al., 1992; Berman et al., 1994; Autor et al., 1998).

Further, there is evidence that in several ways, information technology may be associated with increase in intangible component of output, including variety, consumer convenience, and services. Measuring the intangible component of complementary systems will never be easy, thereby suggesting that computers have made a much larger contribution to the economy than previously believed. This is similar to Stiroh (2002), who also reports a positive correlation between IT capital and subsequent productivity growth across U.S. industries. In particular, industries that made largest investments in computer hardware, software, and telecommunication equipment in 1980s and early 1990s show larger productivity gains after 1995.

The evidence also supports the idea that the acceleration of aggregate productivity is a real phenomenon and not just a cyclical one. The strong and robust correlation between IT intensity and subsequent productivity acceleration, however, implies a deeper relationship between IT investment and productivity growth.

Ferguson and Wascher (2004) review that current productivity boom is associated with a sharp increase in demand for workers with computer-related skills and a further widening in the skill premium for workers with a college degree. As a result, college enrolment rates, which stagnated during the productivity slowdown in the 1970s and 1980s, began to rise again in 1990s. In addition, the demand for lower-skilled workers has been damped in recent years. In response, many lesser-skilled adults have moved to retool their skills by returning to school. Most notably, enrolment at community colleges increased about 30 percent between 1985 and 2000, and the percentage of adults attending an education program raised from 33 percent in 1991 to 45 percent in 1999.

This robust correlation is frequently interpreted as evidence of skill-biased technical change. Autor et al. (2003) formalize and test a simple theory of how the rapid adoption of computer technology, changes the task performed by workers at their jobs and ultimately the demand for human skills. They observe that computer capital substitute for workers in carrying out a "routine tasks". In addition to it, they find that computer capital complement workers in carrying out problem-solving and complex communication activities ("non-routine tasks"). Within industries, occupations, and educated groups, computerization is associated with reduced labour input of routine manual tasks and increased labour input of non-routine tasks.

There is recent literature that shows that internet has changed the markets by allowing more efficient research. Freund and Weinhold (2002) review the scope of growth of trade in services. The international trade in services has been limited (only 20 percent of world trade), as the performance of many services necessitates physical contact between producers and consumers. New technology, in particular, the internet provides a medium of exchange that overcomes such historical trading hurdles for many services, effectively reducing transport costs from infinity to virtually nothing.

The effect of the internet on services trade growth is positive and significant. Countries that have greater internet penetration also have had higher growth in services trade in recent years. The causes probably run both ways: increasing trade in services allows firms to adopt the internet to facilitate the trade, and greater internet penetration causes firms to use internet for trade in services. The results suggest that internet development abroad has already begun facilitating increased export of services to the United States.

The expansion of trade in services has important implications for growth around the world. Some expected standard gains from trade include reallocation of resources to their most productive uses and welfare will increase. Moreover, there are likely to be large multiplier effects, as services are important input into nearly all sectors. This implies that, while internet will affect growth directly through its impact on productivity, it should also increase growth indirectly via its effect on openness.

# 2

# Information and Communications Technology (ICT) in India

India has endeavoured to build and consolidate its knowledge base since independence. Indian planners envisaged the strategy to attain economic and technological self-reliance by investing perceptibly in education, R&D and health sectors. A gigantic network of research laboratories and institutions, coupled with a chain of engineering colleges and IITs has enabled India to forge a formidable reservoir of technically skilled workers. Since the globalization of Indian economy in early 1990s, a massive home-grown technically professionalized manpower gave it strength to emerge as one of the fast growing knowledge economies in the world.

## 2.1 Favourable Conditions for ICT Growth in India

With increase in the use of IT in knowledge intensity sectors of the economy, India has a tremendous growth impetus to emerge as the fourth largest economy in the world (Chadha, 2006). Time is very opportune for India to make its transition to the knowledge economy. India has many of the key ingredients for making this transition. It has a critical mass of skilled, English-speaking knowledge workers, especially in sciences.

It has a well-functioning democracy. Its domestic market is one of the world's largest. It has a large and impressive Diasporas, creating valuable knowledge linkages and networks. In addition, the development of the IT sector in recent years has been remarkable. India has created profitable niches in information technology (IT) and is becoming a global provider of software services. Developing countries are looking at India to find technological solutions for their social problems. Corporations in developed countries are initiating research and development (R&D) activities in India, to develop faster and cheaper products.

India has been witnessing a rapid increase in research and development investments over the past few years. While the government is increasing its budget on research and development each year, more multi-national corporations are initiating research and development activities in India. About 300 multinational corporations have already set up research and development centres in the country, leading to deepening of technological and innovative capabilities among Indian firms.

It is expected that research and development (R&D) investment in India will grow at a compound annual growth rate of 40 percent over the next few years. This will further boost India's ability to carry out world class research. The outsourcing of high end R&D to India is yet another new trend that is evident from large number of established R&D outsourcing centres from IT and telecom sectors in India.

The key factor contributing to India's position as a preferred destination for research and development (R&D) is the availability of a large, highly skilled talent pool. It is the second-largest knowledge resource bank in the world. It has the largest IT pool in the world. There are 1.5 lakh IT professionals in Bangalore. The country has a well-developed higher education system. With 380 universities and 11,200 higher education institutions, India produces over 6,000 Ph.Ds, 2 lakh engineers and 3 lakh science graduates and postgraduates annually. According to World Competitiveness Year Book 2003, around 8 percent of Indian under the age of 25 to 34 years attained tertiary education, as compared to 5 percent in China (FICCI and CSIR, 2006).

After liberalization, foreign direct investment (FDI) flows have been increasing in India. It has created a new competitive advantage for the country especially in the IT domain. With information and communication technologies (ICTs) use becoming all pervasive and its impacts transformational; it has become an essential backbone of the knowledge-economy. India's telecommunication sector has registered rapid growth in recent years, spurred by reforms.

Seeing the growth of Indian software industry, global IT vendors such as Accenture, Electronic Data System (EDS) and IBM Global Services (IBMGS) have rushed to set up their software development centres in India. NASSCOM has reported that growth of these companies is almost twice as fast as Indian companies. The Economist have commented that there has been a shift, albeit slowly, of the geographic location of IT industry from Silicon Valley to new destination such as India (Roy, 2003).

The rise of the software industry has generated a number of externalities like development of entrepreneurship, and reversal of brain drain (Kumar, 2001). Other developmental effects include rise in national income, rise in share of software exports in India's export of goods and services; and increase in employment of software professionals (Verma and Rogers, 2004; Mc Millin, 2006). Chowdhary (2002) reviews that another dynamic aspect of Indian IT industry is IT-Enabled Services (ITES). ITES include services that can be outsourced using powers of IT. The present boom in ITES is primarily in Customer care, followed by administration and banking, all at the lower end of an intellectual pyramid, but involving a large number of people (Gangadhar and Reddy, 2005; Saha, 2005).

According to The Economist (2005a), the entire Indian "Business Process Outsourcing" (BPO) industry, including a wide range of services besides manning the telephone, employs an estimated 3.48 lakh people. This infant industry has grown explosively (Ramesh, 2004). Outsourcing manifests itself in every aspect of business today. The contribution of outsourcing segment to country's gross domestic product (GDP) is expected to be around 8 percent by 2008.

While IT industry in India has emerged as one of the most

dynamic sector over the past decade, the growth of crucial hardware segment of this industry has remained stunted for want of adequate policy support. In the initial years of growth of telecommunication industry, production of hardware was more than software. Nevertheless, after 2000, production of software increased than production of hardware. The total investment that has gone into hardware sector in the country over the last decade is just about ₹ 1,300 crore and because of inability of industry to keep up with the fast changing technologies abroad, some 15 to 20 hardware companies in India either close shop every year or switch over to software development and trading.

The fast growing domestic market for computers and other IT hardware is largely served by imported products with domestic manufacturing playing an insignificant role. According to Azim Premji of Wipro, "If India is to meet the overall export target of US$ 87 billion set for IT sector by 2008; it will require US$ 50 billion worth of hardware". Surely, the country cannot afford to depend on imports entirely to meet their requirement (Naik, 2001).

Because of IT explosion and impressive progress in the telecommunications sector, the usage of ICTs has been growing in the country. However, explosive growth of ICTs has mainly been concentrated in urban areas. As telecom sector moves to a more commercial and competitive environment, the government should implement practical policies to enhance the reach of IT to groups not well served by the market. The challenge is to promote the effective application and use of ICTs throughout the economy to raise productivity and growth, not just in a few pockets.

A knowledge economy is one that creates, disseminates, and uses knowledge to enhance its growth and competitiveness. Successful transition to a knowledge economy is founded on four essential pillars. Making effective use of knowledge in any country requires developing appropriate policies, institutions, and investment and coordination across these four pillars, because of the strong interdependence that exist among them. The economic and institutional regime is in a sense the most critical pillar of the knowledge economy, because it provides the context for the effectiveness of the other three functionally focused pillars (education, innovation, and ICTs).

## 2.2 Four Pillars of Knowledge Economy

These are as under:

1. An economic and institutional regime that provides incentives for the efficient creation, dissemination and use of existing knowledge.
2. An educated and skilled population that can create and use knowledge.
3. An efficient innovation system of firms, research centres, universities, consultants, and other organization that can tap into the growing stock of global knowledge and adapt it to local needs.
4. Dynamic information infrastructure that can facilitate the effective communication and processing of information.

**2.2.1 Economic and Institutional Regime:** The economic and institutional regime allows organizations, people and institutions to adjust to changing opportunities and demands in flexible and innovative way. An effective economic and institutional regime includes a competitive environment that stimulates improved economic performance, a financial system that mobilizes and allocates capital to more productive uses, flexible labour markets. Compared with the rest of the world, India has improved its relative position on this pillar. India leads the South Asia and Africa regions, as well china, on its economic and institutional regime since 1995. Several strengths of India's economic and institutional regime compared with China are democracy; entrepreneurship; stronger infrastructure; supporting private enterprise; efficient capital markets; more advanced legal system and an independent judiciary.

The 1990s have seen progressive integration of Indian economy into global economy. In the early 1990s, liberalization of investment and foreign exchange regime stimulated industrial and service growth. The real depreciation of rupee after the 1990-91 crises promoted exports, and reduction of import barriers allowed more foreign goods into the country. With greater import liberalization and tariff exemptions, Indian firms have gained access to imports at world prices, which has helped to encourage exports.

**2.2.2 Education and Human Resources:** Education is the fundamental enabler of knowledge economy. Well-educated and skilled people are the key to creating, sharing, disseminating, and using knowledge effectively. The knowledge economy of twenty-first century demands ICT competencies in addition to 'hard' skills such as literacy. It includes a new set of skills that include 'soft' skills such as communication skills problem solving skills, creativity and teamwork. These are increasingly important for all workers in the knowledge economy. The development of knowledge economy requires a flexible education system.

It begins with basic education that provides for foundation of learning; continues with secondary and tertiary education that develops core skills; encourages creative and critical thinking that is the key to problem solving and innovation; and extends into a system of lifelong learning. A large pool of highly educated and vocationally qualified people in India is making their mark at home and abroad in science, engineering, IT, and R&D. However, they make up only a small fraction of India's population. To become a knowledge economy India needs to keep developing its human capital base creating knowledge workers to contribute to its growth, development and competitiveness in the global economy.

Since Independence, India's economic development has hinged on the strategy to achieve economic and technical self-reliance. It was evidently visualized that self- reliance would be unattainable unless based on domestically groomed work force and skill. A strong emphasis was laid on an accelerated rate of human capital formation by perceptibly investing in education, health and research and training. In its size and diversity, India has the third largest higher education system in the world, next only to China and the United States. Before Independence, access to higher education was very limited and elitist, with enrolment of less than a million students in 500 colleges and 20 universities.

Since Independence, the growth has been very impressive; the number of universities has increased by 18 times from 27 in 1950 to 504 in 2009. The number of colleges has also increased manifold from 578 in 1950 to more than 30,000 in 2011. The system is now more mass based and democratized with one third to 40 percent of enrolments coming from lower socio-economic strata, and women comprising of some 35 percent of the total enrolments. Education in India has received enormous financial support from both the Central and the State Governments. In 1962, the total public expenditure on education was only 1.52 percent of GDP and in 2009 it has grown to 3.78 percent of GDP as shown in Table 2.1. An analysis of total expenditure on health shows that the share of total expenditure on health in GDP has increased to 6.10 percent in 2009 as compared to 5.10 percent in 1998.

Recent developments have shown that India has become a major international centre for the recruitment of high-quality IT staff. Many IT workers leave, but many return, and both flows generally lead directly and indirectly to significant knowledge transfers and linkages across businesses entities that benefit Indian society as well as individuals. Young IT professionals as well as medical professionals want to go abroad mainly to gain professional experience, which they think will be highly valuable in India when they come back.

In addition, they are encouraged by higher earnings, perks, and high quality of life in the host country. Bangalore is becoming a 'corridor' for migration (outward and inward) of Indian human resources in sciences and technology. IT professional in Bangalore believe they have growing opportunities for career growth in India in general and Bangalore in particular. India must keep on building a cadre of technicians, professionals, and knowledge workers who will be the backbone of the knowledge-based economy.

**Table 2.1: Share of Expenditure on Education and Health in GDP (Percent)**

| Years | Share of Public Expenditure on Education in GDP | Share of Total Expenditure on Health in GDP |
|---|---|---|
| 1962 | 1.52 | - |
| 1981 | 2.98 | - |
| 1998 | 3.49 | 5.10 |
| 2000 | 4.19 | 5.00 |
| 2001 | 4.28 | 4.90 |
| 2002 | 3.81 | 4.80 |
| 2003 | 3.78 | 4.10 |
| 2004 | 3.51 | 4.10 |
| 2005 | 3.36 | 4.20 |
| 2006 | 3.46 | 4.10 |
| 2007 | 3.64 | 4.10 |
| 2009 | 3.78 | 6.10 |

Source: http://www.who.int/whr/2006/annex/06_annex2_en.pdf
http://education.nic.in/planbudget/GDP51-09.pdf

**2.2.3 Innovation System:** The third pillar of the knowledge economy is the existence of an efficient innovation system. The innovation system plays an important role in acquiring, creating, adapting and disseminating knowledge. The innovation system in any country consists of network of institution, rules and procedures that affect creation and dissemination of the knowledge. For a developing country like India, tapping and effective use of existing global knowledge will have a greater economic impact.

India lags behind China in spending on research and development work as well as number of scientific researchers.

Number of persons doing research and development in Scandinavian countries was 7,000 per million of population and 4,700 per million of population in the US while in India, there were 156 researchers per million of population in 2008. India is placed within the lower end of dynamic technology adopters. It is because of huge variations in technological achievements among Indian states. Other reasons also exist.

Even though the country has one of the world's largest pools of scientists and engineers, its average years of schooling were just more than five years and the adult literacy rate was about 65 percent. These indicators matter, because they affect the ability of a country to harness technology and innovations for its developments. To its credit, India has been making notable progress in terms of strengthening R&D infrastructure, developing technological innovations and alerting the mindset of its people towards better technology absorption.

Despite the challenges, the overall Indian innovation system presents some remarkable features. India has a dynamic hub of innovation: Bangalore, with its 1.50 lakh strong army of software engineers has a developed a strong innovation "cluster" in IT, characterized by corporate offices, venture capitalists, and business start- ups, and university and research labs. India also has vast and diversified publicly funded R&D institutions such as Indian Space Research Organization, CSIR, Indian Council of Medical Research, and World Class higher education Institutions (IITs and RECs). These institutions provide critical human capital needed for the innovation.

India has recently been very successful in attracting R&D investments. About 100 MNCs have set up R&D centres in India, for example, GE, McDonald's, General Motors, Delphi, Eli Lilly, Hewlett-Packard etc. Large companies are coming to India to take not only the advantage of India's human resource, but also to reap cost advantages. Not all MNCs in India, however, work at cutting edge of technology. Most of them do developmental research, which develops or improves existing products. Some have expressed concern that the trend to MNC R&D investment in India is exploiting not only its most valuable human resource but it has important benefits for Indian industry. Linkages in R&D between local universities, research institutes, and firms on the one hand, and the R&D network of multinationals on the other will help to integrate India further into global technology development activities.

The new trend of outsourcing of high-end R&D to India is evident from the large number of established R&D outsourcing centres in India, from IT and telecom to automotive and pharmaceutical sectors. The R&D outsourcing markets for IT industry in India is expected to grow from US$ 1.30 billion in 2003 to more than US$ 8 billion by 2010 (NASSCOM, 2008). Outsourcing in new areas by companies from developed countries is gradually leading to the development of many critical skills and technologies needed in the country. It will help to raise India's technological capability with relatively modest investments.

**2.2.4 Information Infrastructure:** Fourth pillar of knowledge economy is a dynamic information infrastructure that can facilitate effective communication, dissemination and processing of information. Rapid advances in ICTs are dramatically affecting the acquisition, creation, dissemination and use of knowledge. It affects economic and social activities including how manufacturers, service providers and government are organized and how they perform these functions. Applications of ICTs are improving the efficiency of existing services and creating new opportunities in trade, governance, education, business activity, health-care delivery and environmental and natural resource development. The use of ICT is reducing transaction cost, time and space barriers. It allows the mass production of customized goods and services and it has become a backbone of the knowledge economy. This means that countries need to harness to full potential of ICTs for all sectors of economy: for education, innovation, learning, public and private sector management.

The information infrastructure in a country consists of telecommunication networks, information systems, policy framework and skilled human resource. The government of India created favourable conditions to increase ICT penetration in the country. India's telecoms sector has registered rapid growth in recent years due to reforms and introduction of greater competitions in the sector. The number of fixed lines and cellular subscribers in India has increased perceptibly, especially in the post-reform period. The rapid strides in the telecom sector have been facilitated by liberal policies of the Government that provides easy market access for telecom equipment and a fair regulatory framework for offering telecom services to the Indian consumers at affordable prices.

According to TRAI, the wireless subscribers are not only more than wire line subscribers in the country, but also increasing at a much faster pace. The share of wireless phones in the total subscribers has increased from 2.28 percent in 1996-97 to 96.61 percent in 2011-12. This growth has resulted in increase in teledensity in India to 78.66 as on March 31, 2012.

Despite these achievements, when viewed in a global perspective, the rates of teledensity and ICT penetration for India remain rather low as depicted in Table 2.2.

The global bench marking shows that even though India does better than South Asia and Africa Regions in terms of Information infrastructure indicators, the country has slightly worsened its performance between 1995 and 2005, whereas China has moved far ahead. This is because, despite the several-fold increase in number of fixed line telephones, computers and internet users in absolute terms; the country has maintained its position in telephones, computers but fallen behind in internet users in relative terms. China has made a remarkable progress in improving its overall infrastructure. To take an example of just one indicator; total ICT expenditure as percentage of GDP in India was 2.78 percent in 2002, compared with more than double at 5.81 in China and 8.3

percent in Brazil.

The low level of spending in India has been due to a high level of regulation, lack of local applications, and high tariff on hardware, among others. More investment is required so that India can make strides in strengthening its infrastructure.

**Table 2.2: ICT Indicators for India and China, Various Years**

| Variable | India | China |
|---|---|---|
| Telephone subscribers per 100 inhabitants, 2007 | 23.36 | 68.71 |
| Mobile subscribers per 100 inhabitants, 2007 | 19.98 | 41.19 |
| Internet users per 100 inhabitants, 2007 | 6.93 | 16.10 |
| Computers per 1000 people, 2002 | 7.20 | 27.60 |
| T.V. sets per 1000 people, 2002 | 83.00 | 350.00 |
| Radios per 1000 people, 2001 | 120.00 | 339.00 |
| Daily newspapers per 1000 people, 2000 | 60.00 | 42.00 |
| Internet host per 1000 people, 2003 | 0.82 | 1.28 |
| Internet users per 1000 people, 2003 | 174.86 | 632.48 |
| International telecommunication, cost of call to the US (US$ per 3 minutes), 2002 | 3.20 | 6.70 |
| ICT expenditure as percent of GDP, 2002 | 2.78 | 5.81 |

Source: (i) Dahlamn, C. and A. Utz (2005), "*India and the Knowledge Economy: Leveraging Strengths and Opportunities*" WBI development Studies, Washington D.C.: The World Bank. (ii) International Telecommunication Union (2009), *Basic Indicators*, accessed on 26th April 2009, available at http://www.itu.int/ITUD/icteye/Indicators/Indicators.aspx#

## 2.3 India's Global Standing in Information Technology

A number of indexes have been developed to rank countries' performance on ITs. The indexes provide a variety of ranking on where India stands in a global perspective and differ with regard to their context and specific informational functions. One possible explanation for the differences in India's ranking is the way the different indexes are calculated; all of them use different methodology and differ in scope in terms of the number of countries and indicators covered.

**2.3.1 Networked Readiness Index:** The World Economic Forum's Global Information Technology Report of 2004-05, highlights its Networked Readiness Index (NRI) of 104 countries. The NRI measures the degree of preparation a nation or community has to participate in and benefit from ICT developments. The NRI is composed of three components:

1. Environment for ICT offered by a given country or community.
2. Readiness of the community's key stakeholders.
3. Usage of ICT among these stakeholders.

The NRI for 2004-05 places Singapore as the best performer worldwide. Singapore is followed by Iceland, Finland, and Denmark, and the United States occupies the fifth place. In terms of India and its comparators, Korea tops the list and holds 24th place, but India is next on the list at 39th, surpassing China at 41st, Brazil at 46th, and Russia at 62nd.

**2.3.2 E-readiness Ranking:** It provides another interesting benchmark for countries in comparing and accessing their IT related environment. E-readiness takes into account factors like quality of IT infrastructure, government initiatives and the degree to which internet is creating real commercial efficiencies. The ranking covers world's largest 64 economies. In the 2004 ranking, several countries have improved their scores. Four tires of E-readiness have emerged in 2004 ranking:

1. Denmark holds the top spot, and Korea is ranked 14th. Korea is the densest broadband market in the world.
2. Second tire includes countries like Brazil (35th) and Poland (36th) that do not have dense communication internet infrastructure.
3. Third tire includes large 'e-ready' economic powerhouses: India (46th), China (57th), Russia (55th).

4. In the fourth tire is a cluster of markets in which internet services are struggling.

**2.3.3 Information Society Index:** The Information Society index (ISI) in 2004 measured the abilities of 53 nations to participate in information revolution. The ISI combines 15 variables in four infrastructure pillars. The four pillars are:

1. **Computers:** This pillar look at basic building blocks of information society by measuring the PC households, IT spending as percentage of GDP, software spending as percentage of total IT spending and IT services spending.
2. **Internet:** The internet is the key factor in development of advanced information society. It takes into account number of internet users, percentage of internet users at home, number of mobile internet users at home and e-commerce spending at home.
3. **Telecoms:** This pillars measures broadband adoption, wireless services and mobile handset shipments.
4. **Social:** This pillar evaluates a society's ability to utilize information technology by measuring education, civil liberties and government corruption.

In 2004 ranking, Denmark was in first place, followed by Sweden, Korea (8th). Brazil took 38th place, Russia 41st, China 44th and India 51st.

**2.3.4 E-readiness Assessment of Indian States:** According to this report India differs from others by looking at the level of e-readiness of various Indian states. The report classifies Indian states into the following categories:

1. **Leaders:** Karnataka, Tamil Nadu, Andhra Pradesh, Maharashtra and Chandigarh.
2. **Aspiring Leaders:** Kerala, Gujarat, Delhi, Punjab and Haryana.
3. **Expectants:** West Bengal, Pondicherry and Madhya Pradesh.
4. **Average Achievers:** Uttar Pradesh, Chattisgarh, Orissa, Sikkim, Himachal Pradesh and Rajasthan.
5. **Below Average Achievers:** Mizoram, Jammu and

Kashmir, Assam, Meghalaya, Uttrakhand and Jharkhand.

6. **Least achiever:** Lakshadweep, Manipur, Tripura, Arunachal Pradesh, Andaman Nicobar Islands, Bihar, Daman and Diu and Nagaland.

IT infrastructure and the ability to use it are clearly dispersed across Indian states. Some states in India, such as Tamil Nadu have improved by involving private sectors in development of IT infrastructure and introducing ITs in State-level policies. Kerala has used the high proportion of literacy and awareness of citizens for the development of ITs. Madhya Pradesh has developed largely due to private sector involvement in developmental activities. In all cases, these Indian states have made progress due to leadership, connectivity, availability of skilled workers, increased private sector development and creation of institutional mechanism.

It is clear that developing countries must invest in building their information infrastructure; this includes not only telecommunication services, but also ensuring that these services are provided at the lowest cost. They should promote the ITs to facilitate improvement in economic and social activities. Furthermore, as the applications of the ITs becomes more important for economic growth and social welfare, special attention must be paid to new investment in improving the skills and competencies of workforce to use ITs.

## 2.4 Market Concentration of Indian IT Industry

Information Technology industry can be broadly classified into two broad segments-computer software and computer hardware. The market for computer software and computer hardware is highly competitive. The Herfindahl Index (H-Index) for computer software and computer hardware works out to be 0.102 and 0.198 in 2006-07 respectively. Smaller value of H-Index indicates more dispersement in the concentration of market share.

In the initial phases of development of software sector, value of H-Index declined to 0.045 in 2000-01 showing decrease in market concentration. H-Index has been increasing since 2000-01 and in 2006-07 it stood at 0.102, thereby showing increase in market share for few firms. In late 1990s, significant policy initiatives were introduced by Government in the direction of privatization in telecommunication sector. These policy reforms opened new opportunities for domestic IT firms and number of firms grew significantly in 2000 and the year 2000 marked a structural break in the industry (Kumar, 2001).

According to U.S. Government estimates, more than 260 of fortune 1000 companies outsourced their software development to India in the year 2000 (Kumar, 2001; Aggarwal, 2009). Similar tendency is found in hardware sector. The H-Index for hardware sector declined from 0.092 in 1991-92 to 0.072 in 2000-01, showing decrease in market concentration. Though it increased to 0.198 in 2006-07 showing rise in market share for few hardware companies, yet the concentration is quite low.

However, the IT industry is highly fragmented, with top 10 firms accounting for less than 20 percent of the total industry revenue in 2001 (NASSCOM, 2003). The IT industry comprises several companies with varying sizes and engaged in one or more specialized activities. The fragmentation of IT industry is primarily due to the following factors: (a) wide range of software products, hardware and technology standards; (b) different needs of companies, and (c) there are low entry barriers for new entrants in the value added segment of the industry.

**2.5 Size Distribution of Indian Software-Export Industry**

Another attribute of the concentration is the relative sizes of firms measured by their market share. It influences the behaviour of the firms in the industry/market. In case of inequality in the size of firms, fluctuations in the output of large firms will significantly affect industry output and hence price. The behaviour of small firms is likely to have a negligible effect on the market share of other firms, while a large firm trying to expand can do so by significantly affecting the others. Bishop (1952) concludes that the size distribution will have perceptible effects, independent of the number of firms. Mancke (1974) finds that firm profitability is correlated with market share and size.

The Indian IT industry has different size groups of companies- large, near billion dollar (above ₹ 1000 crore) companies, small start-ups, Indian companies and multinationals. These different groups of firms have different growth rates.

Indian IT industry is dominated by a few large firms but there are a large number of small and medium firms also. Top 20 firms accounted for 46 percent of total revenue of IT industry in 2002-03 and their share has grown over time. In 2008-09, the share of top 20 firms stood at 64 percent. Software Technology Park of India (STPI) has facilitated the promotion of long tail industrial structure by promoting Small and Medium Enterprises (SMEs).

These firms have the potential to offer niche and super specialized work (Gupta, 2006). The total number of firms registered with STPI is more than 9700 (Aggarwal, 2009). However, more than one third of revenue generated by the industry is accounted by the 12 largest companies. These are Infosys, Wipro, Tata Consultancy, Satyam, Mphasis, Patni, Cognizant, Polaris, Tech Mohindra, Hexaware, Mindtru and HCL. The rest of firms were small and medium sized.

Currently, IT industry is witnessing fundamental structural changes. Initially, companies chose their core areas like hardware, software, services, consultancy, outsourcing, search engine etc. Now they are venturing into other territories, thereby widening their scope, like Microsoft entering into search engine territory; Google entering software territory, and IBM selling PC business to China's Lenovo.

The broad based structural changes in the industry can be attributed to componentization, convergence of different technologies especially computing and communications, stronger linkages between business process and supporting technologies, and delivering software as a service (SAAS) etc. In other words, the reasons truly span from raw technology innovation to evolving delivery models. As a result, the market structure of the Indian IT sector has become quite broad-based reflecting its relative maturity over time.

**2.6 Market Structure of Hardware Segment**

The Indian domestic market has recorded a smaller base and lower growth rate as compared to the export segment. The domestic market has grown significantly since 1993 and gradually the industry has shown an upward trend in growth. A strong domestic market is always an asset. It fuels growth for industries in the global market, helps them to compete effectively, supports rapid accumulation of skills and offers an insight into product and process needs (Government of India, 2002b).

Hardware accounts for majority share of the domestic market in India. The market structure of domestic hardware segment reflects dominance of MNCs (Rao, 2009). The MNCs' share in the key segments of hardware components ranged 60 percent to 95 percent in 2003-04. In the servers' category, HP, IBM and Sun occupied the top three positions. The domestic PC market showed signs of maturity as branded PC firms managed to gain share from assemblers.

HCL Info systems managed to retain its leadership in desktop PCs, followed by three MNCs firm HP, IBM, and Dell in the list of top five. On the other hand, notebooks segment is dominated by MNCs (approx. 95 percent) with HP, IBM, Toshiba, Acer and Dell competing for the top five lists in the segment. In peripherals category (printers, camera, keyboards, optical mouse, optical disc drivers etc.) market leader HP created multifunctional devices followed by Canon and Lexmark in the top three lists in the segment.

Computer hardware is an essential requirement for smooth

production of IT and ITES output. The hardware not only has to be available in sufficient volume at cheap rates but it has to be the latest in technical innovation. Over the years, the ownership profile of computer hardware industry has changed. Broadly, there are four phases of its evolution. The first phase was marked by foreign domination followed by a period where public sector controlled the industry. During the third phase, the private sector firms began to dominate and in the final phase, which coincided with the globalization phase, the foreign firms again had taken control over the industry (Joseph, 1997).

Domestic market for IT services is witnessing a gradual transformation from being predominantly hardware driven towards a solution-oriented approach. It has resulted in growing emphasis on services. A host of factors such as global slowdown impacting overall IT spending by enterprises, fluctuating dollar prices and strengthening of the Indian rupee against dollar led companies like IBM, HP, TCS, and Wipro InfoTech to venture deeper into the domestic IT services market. The domestic IT sector has been growing at a decent rate.

The market segments that are expected to witness the highest growth is infrastructure services and enterprise IT outsourcing. Infrastructure services play a critical role in business transformation, which includes mainframe centric solutions, desktop and distributed computing, network operations and monitoring, asset management, service delivery management, e-commerce and collaborative computing. On the other hand, growth of enterprise IT outsourcing in India is expected to fuel the surge of IT consultancy services, which occupies just 5 percent of the overall market in 2007-08.

## 2.7 Location Pattern of Indian IT Industry

Since its perceptible growth in early 1990s, like most of productive investments, IT investments were pulled naturally too few centres of agglomeration, which offered various economic advantages. In recent times, many studies like Scott (1996); Dickson (2003); Hutchison and Ilavarason (2008) have tried to empirically analyze the location decisions by firms. A large part of India's IT industry growth has concentrated in about 5-6 urban centres or tier-I cities such as Bangalore, Mumbai, Hyderabad, Delhi (including Noida and Gurgaon or NCR) and Chennai. Bangalore, where many of IT firms are located, has been rated 11th among the 46 top global hubs of innovation (UNDP, 2001).

The highly influential Diasporas of Indian technologists have played a key role in bringing high technology investment to Bangalore, Hyderabad and other Indian cities (The Economist, 2005b). Table 2.3 shows the dominance of five clusters namely Bangalore, Chennai, Mumbai, Hyderabad and Delhi. These five cities account for 80.5 percent of the top 600 firms in 2002. The high value of Ginni Coefficient shows that distribution of firms is skewed towards few areas.

A number of factors determine the attractiveness of a location, like availability of human resource (number of employees in IT industry, average salary and attrition rate); infrastructure (rental rates, telecom connectivity and power); ITES-BPO focus (number of ITES-BPO companies, nature of work and policy); support group presence (local support, level of participation and knowledge base generated); and physical resources(geographic location, climate conditions and availability) (NASSCOM, 2005; Rai, 2005; Ribeiro, 2005; Hutchison and Ilavarason, 2008; Bhatt and Illiyan, 2009).

The Government of Karnataka has proclaimed the IT sector as a core sector going forward. Bangalore continues to promote IT industry with over 30 SEZs and a number of new IT parks. Bangalore is India's fifth largest metropolitan area. There are a large number of public sector heavy industries for example, aerospace, telecommunication, machine tools, heavy equipment and defence establishments, in addition to biotechnology, IT and BPO industry. In terms of availability of human resource, city has around 75-90 colleges producing around 20 thousand engineering graduates per year and Infosys, Wipro, Satyam, TCS, IBM, Oracle, Lucent are some of top recruiters in IT services.

**Table 2.3: Geographic Location of IT Industry (2002 and 2009)**

| Region | Location of Top 600 Firms in 2002 (percent) | Location of ITES Firms | | Location of 1237 Firms as on December 2009 (percent) |
|---|---|---|---|---|
| | | Number | Percentage | |
| Mumbai | 21.8 | 45 | 16.2 | 16 |
| Bangalore | 20.3 | 35 | 12.6 | 22 |
| NCR | 18.5 | 53 | 19.1 | 22 |
| Hyderabad | 10.7 | 24 | 8.6 | 9 |
| Chennai | 9.2 | 35 | 12.6 | 13 |
| Kolkata | 4.2 | 29 | 10.4 | 3 |
| Pune | 3.8 | 6 | 2.2 | 6 |
| Trivandrum | 2.3 | n.a. | n.a. | _ |
| Kotchi | n.a. | 10 | 3.6 | _ |
| Ahmedabad | n.a. | 9 | 3.2 | _ |
| Others | 9.2 | 32 | 11.5 | 9 |
| Ginni Coefficient | -- | 0.70 | – | – |

Note: n.a.- not available.

Source: (i) NASSCOM (2003), *Strategic Review: IT Industry in India,* New Delhi: NASSCOM (ii) NASSCOM (2010), *Annual Report 2009-10,* New Delhi: NASSCOM.

Around 30 thousand graduates with good proficiency in English and Hindi skills and technical training are available for BPO sector. The city is well connected to the rest of country by various modes like rail, road, and air. A new airport has been constructed to cater to the increasing demands for air travel. In addition to good education, Bangalore has adequate recreational facilities, excellent health care facilities and few power cuts for short duration. The city is ranked 14th on crime index (median rate of crime) and fifth on the cost of living index (NASSCOM-A.T. Kearney, 2008).

Similarly, Chennai is a large commercial and industrial centre with main industries being automotive, software services, BPO, hardware manufacturing and financial services. The state government has encouraged the growth of IT-BPO industry and Chennai ranks amongst the top three as per the 2007 Investment Environment Index. Chennai provides one of the largest out-turns of engineers in the country i.e., around 35 thousand engineering graduates and 40 thousand graduates annually. National and international major IT firms have been recruiting engineering graduates from the campuses for both local and global offices.

The availability of large number of chartered accountants makes it an ideal destination for IT-BPO companies offering services in the financial sector. The city has excellent power supply and national and international connectivity. There are large number of five star hotels in the city and over three thousand hotel rooms(in various star category) are expected to come up by October 2008-09. The city has got good medical facilities and is ranked 28th on the crime index.

In a similar vein, NCR has a large base of IT companies that recruit regularly from colleges. Graduates out-turn across NCR is expected to increase by more than 60 percent over the next two years because of establishment of new colleges. With IT being the preferred sector for engineering students, a sizeable resource pool for IT exist which can enhance the growth of IT industry. Delhi is an established IT-BPO destination with Location Attractiveness Score of 6.1 and 6.3 for IT services and BPO services respectively. It has approximately 60 percent cost advantage in comparison to the average cost of leader locations (NASSCOM-A.T. Kearney, 2008) and the city has the potential to support IT-BPO firms in excess of current levels.

Thus, first-tier locations will continue to be the prime locations for IT/ITES operations in the years to come. This is due to the factors such as their well-developed labour markets, established lobby groups, and developed supporting industries. Markusen (1995) argues, "Because high-tech industries are youthful, reliant on external services, and subject to rapid rates of innovation, they tend to locate in the major centres of innovation and diversified economic activity where agglomeration economies can be realized".

Software Technology Park of India (STPI) has played a stellar role in facilitating the geographic expansion of the industry with in the country. One of the objectives of STPI in the initial phases was to provide data communication facilities to software exporters. In persuasion of this objective STPI established its own gateways at its nodal centre located in different parts of the country and maintained internal engineering resources to provide consulting, training and implementation services.

In the late 1990s, significant policy initiatives were introduced by the government in the direction of privatization in telecommunication. These telecommunication policy reforms opened new opportunities for domestic firms. In 2005, The Government of India facilitated availability of bandwidth at low rates due to competition among a number of bandwidth providers, cheap optical fiber connectivity against the satellite connectivity and duty-free imports.

In this scenario, the role of STPI in providing these services has become relatively less important. However, this has not diminished the importance of the STPI scheme in the promotion of the IT industry.

Units in STPs are benefited by the single window approval mechanism, availability of commercial infrastructure at competitive rates, incubation facilities and other infrastructure. The STP also offers a transparent policy environment and a package of concessions. It thus helps not only in enhancing cost competitiveness by lowering cost of production but also ensures conducive business environment.

## 2.8 Public Sector versus Private Sector

The ownership structure of the telecommunication sector can be mainly divided into public sector and private sector service providers.

Among the public telecom operators are Bharat Sanchar Nigam Limited (BSNL) and Mahanagar Telephone Nigam Limited (MTNL) while among private telecom operators are Reliance Telecom, Bharti Infotel, Tata Teleservices, Shyam Telelinks, Bharti Cellular, Bharti Airtel, Idea Cellular, Vodafone Essar, Aircel etc.

Intensive competition among few private sector operators contributed prominently to post-1998 growth. Public sector operators expanded their subscriber base by 32 million during 1998-2006 whereas the private sector subscribers grew by 80 million during the same period.

It is remarkable that public sector telecommunication services whose growth was slow during the monopoly period of 1948 to 1998, i.e., addition of about 0.3 million subscribers every year, inflated to the addition of about 5 million subscribers every year during 1998 to 2005.

The public-private co-operation and competition led to immense improvement in the performance of both public and private sector service providers.

Companies behaved very differently in monopoly and competitive environments and consumers only gain when competition is introduced in markets.

The share of private sector in total telephone connection has steadily increased to 72.40 percent in December 2007 as against 2.28 percent in 1996-97 (Government of India, 2008). Table 2.4 depicts that the share of public sector in total revenue of telecom services declined from 71.13 percent in 2001-02 to 28.57 percent in 2008-09.

This is due to the fact that the public sector is more dominant in wire line or fixed services and the private sector dominates in the wireless or mobile services segment.

BSNL and MTNL are the two public sector undertakings

that meet the growing requirements of telephone and other related services. Among private players, top 5 include Tata/Hughes, Bharti, Reliance, HFCL and Shyam. In addition to it, there are five licensed private operators providing basic services. These are Reliance Infocom (operating in 18 service areas), Tata Teleservices (Maharashtra, Andhra Pradesh, Tamil Nadu, Chennai, Karnataka, Gujarat), Bharti (Madhya Pradesh, Karnataka, Haryana, Delhi, Tamil Nadu), HFCL (Punjab) and Shyam Telelinks (Rajasthan).

**Table 2.4: Share of Public and Private Sector in Total Revenue of Telecom Services (Percent)**

| Years | Public Sector | Private Sector |
|---|---|---|
| 2001-02 | 71.13 | 28.87 |
| 2002-03 | 69.37 | 30.63 |
| 2003-04 | 63.01 | 36.99 |
| 2004-05 | 56.31 | 43.69 |
| 2005-06 | 53.78 | 46.22 |
| 2006-07 | 37.64 | 62.36 |
| 2007-08 | 30.10 | 69.90 |
| 2008-09 | 28.57 | 71.43 |

Source: (i) CMIE (various issues), *Financial Aggregates and Ratios* (various issues), Mumbai: CMIE, Economic Intelligence Service. (ii) http://voice data.ciol.com/content/stats/default.asp.

## 2.9 Market Share of Internet Services Operators

The internet services in India have developed at high speed and with wide coverage. By the end of March 1998, the number of internet subscribers stood at 1.40 lakh. In November 1998, the government recognized the need for encouraging the spread of internet in the country and opened the sector for private operators and thus ended the monopoly of internet services by public sector VSNL. The Government of India liberalized the terms and conditions of internet service providers (ISPs) licenses with no license fee and allowed unlimited number of providers. The ISPs could set their own tariffs and even their own international gateways. The Department of Telecommunication (DoT) issue three type of licenses-Category 'A' for all India operations; Category 'B' for metros and state-level circles and Category 'C' for medium and small cities.

The Indian Government has allowed foreign companies to hold 49 percent share of ISPs. In 2007, there were 389 ISP licensees, but 135 were operational. Due to increased competition among few ISPs, H-Index plunged sharply. It nose-dived from 0.224 in 199-00 to 0.136 in 2006-07 showing a much-diffused pattern of market share among few ISPs. Internet communication in India is developing rapidly and public sector providers (BSNL/MTNL) dominated the market with 62 percent of the total internet subscriber base in 2006 as shown in Table 2.5. As on March 31, 2012, the share of BSNL/MTNL stood at 72.2 percent. Their dominance in fixed landline services put them in an advantageous position to offer combo plans (where user of voice service can avail internet service also). The total broadband subscribers in India have reached 13.79 million in 2011-12, as compared to 3.9 million in 2007-08. At present, the broadband market is dominated by BSNL, Bharti Airtel, and MTNL.

**Table 2.5: Market Share of Top Five Internet Service Providers**

| Internet Service Provider | Market Share ( percent) | |
|---|---|---|
| | 2005 | 2006 |
| BSNL | 34.24 | 43.07 |
| MTNL | 18.85 | 19.26 |
| Sify Ltd. | 14.36 | 11.83 |
| VSNL | 10.88 | 6.12 |
| Bharti Televentures | 4.82 | 5.73 |
| Total | 83.15 | 86.01 |

Source: (i) Government of India (2006), *Analysis of Internet and Broadband tariffs in India*, New Delhi: TRAI, Study Paper No. 2/2006. (ii) Government of India (2005), *The Indian Telecom Services-Performance Indicators,* New Delhi: TRAI, April-June.

While BSNL is the leader in the broadband market based on subscribers, Reliance Communications is the leading player based on revenues.

As India increasingly adopts an online environment, more and more services are being made available online. Thus, e-commerce is playing an important role in driving the growth of broadband services in India. The other factors include faster download speeds, lower costs of equipment (like computers) to enable broadband access, increasing downloads of content (like music and movies) from the Internet, and other value added services (like gaming) on broadband, etc. Bharti Airtel has gone a step further to popularize broadband amongst the youth by launching games on demand to cater to the youth besides the introduction of innovative services (like video surveillance), which run on a broadband environment.

Despite the liberal policy of Government of India, the growth of internet subscribers has not followed the same pattern as the number of telephone connections, although the number of internet users has been rising relatively faster. It may be due to the concentrated nature of the market for internet service providers (Mani, 2008).

In nutshell we find that the software exporting industry has emerged in India in 1974 under the highly protected environment. The rapid growth of global demand for software pointed to increasing export potential which in turn called for more concrete policies for promotion of software development and exports. Accordingly, in 1986, an explicit software policy was announced. In addition to it, Government of India started telecom reforms since1984 which played a key role in the growth of sector. Software Technology Park also helped not only in enhancing cost competitiveness by lowering cost of production but also in providing conducive business environment.

The structure of IT industry reveals that the market for computer software and computer hardware is highly competitive. Smaller value of The Herfindahl Index (H- Index) for computer software and computer hardware industry indicates dispersion of market shares. The market for mobile services is much more competitive than the one for fixed line services. The ownership structure of the telecommunication sector comprises public sector and private sector service providers. The public sector is more dominant in wire line or fixed services and the private sector dominates in the wireless or mobile services segment. Despite the liberal policy of Government of India, the growth of internet subscribers has not followed the same pattern as the number of telephone connections, although the number of internet users has been rising relatively faster. It may be due to the concentrated nature of the market for internet service providers

It has been observed that the major part of India's IT industry has remained concentrated in about 5-6 urban centres or tier-I cities such that Bangalore, Mumbai, Hyderabad, Delhi (including Noida and Gurgaon or NCR) and Chennai. The high value of Ginni Coefficient shows that distribution of firms is skewed towards few areas.

# 3

# Legal and Institutional Framework for ICT

India has emerged in recent years as the most preferred destination for outsourcing of IT services. The vibrant IT industry is contributing immensely by providing information about latest technology and international business practices. Various sectors of the Indian economy such as industry, finance, insurance, communications and transport have adopted information technology in a big way. However, agricultural sector of the economy is lagging behind in utilizing IT services. Some efforts have been made in this regard but they are by no means adequate.

## 3.1 Information Technology Act (ITA), 2000

On January 30, 1997, the United Nations General Assembly adopted, through a Resolution, the Model Law on Electronic Commerce adopted by the United Nations Commission on International Trade Law. This is referred to as the UNCITRAL Model Law on E-Commerce.

Following the UN Resolution, India Parliament passed the Information Technology Act (ITA), 2000 in May 2000 and notified it for effectiveness on October 17, 2000.

ITA, 2000 consisted of 94 sections segregated into 13 chapters. Four schedules also formed part of the Act.

ITA, 2000 addressed the following issues:

1. Legal Recognition of Electronic Documents.
2. Legal Recognition of Digital Signatures.
3. Offences and Contraventions.
4. Justice Dispensation System for Cyber crimes.

ITA, 2000 was substantially amended through the Information Technology Amendment Act, 2008 which was passed by the two houses of the Indian Parliament on December 23 and 24, 2008. It got the Presidential assent on February 5, 2009 and was notified for effectiveness on October 27, 2009.

In the 2008 version of the Act, there are 124 sections (excluding 5 sections that have been omitted from the earlier version) and 14 chapters. Schedule I and II have been replaced. Schedules III and IV are deleted.

ITA, 2008 as the new version of ITA, 2000 has provided additional focus on Information Security. It has added several new sections on offences including Cyber Terrorism and Data Protection.

The amendment was passed in an eventful Parliamentary session on December 23, 2008 with no discussion in the House.

Some of the cyber law observers criticised the amendments on the ground of lack of legal and procedural safeguards to prevent violation of civil liberties of Indians. There has also been appreciation about the amendments from many observers because it addresses the issue of cyber security.

### 3.2 Department of Information Technology (DIT)

DIT is a Department of the Ministry of Communications and Information Technology, Government of India. It performs the following functions:

1. Policy matters relating to information technology, electronics and internet.
2. Initiatives for development of hardware/software industry including knowledge-based enterprises, measures for promoting information technology exports and competitiveness of the industry.
3. Promotion of information technology and information technology-enabled services and internet.
4. Assistance to other departments in the promotion of e-governance, e-infrastructure, e-medicine, e-commerce etc.

5. Promotion of information technology education and information technology-based education.
6. Matters relating to cyber laws, administration of the Information Technology Act, 2000 and other information technology related laws.
7. Matters relating to promotion and manufacturing of semiconductor devices in the country.
8. Interaction in information technology related matters with international agencies and bodies.
9. Initiative on bridging the digital divide, matters relating to Media Lab Asia.
10. Promotion of standardization, testing and quality in information technology and standardization of procedure for information technology application and tasks.
11. Electronics Export and Computer Software Promotion Council (ESC).
12. National Informatics Centre (NIC).
13. All matters relating to personnel under the control of the Department.

## 3.3 Organizations of Department of Information Technology (DIT)

**3.3.1 Controller of Certifying Authorities (CCA):** Information Technology (IT) Act, 2000 was enacted by the Indian Parliament in June 2000. It was notified for implementation in October 2000 with the issuance of Rules under the Act. The purpose of the Act is to promote the use of digital signatures for the growth of E-commerce and E-governance. CCA is the authority under which Cyber Regulations Appellate Tribunal (CRAT) has been established under the IT Act. See CRAT below.

**3.3.2 Cyber Regulations Appellate Tribunal (CRAT):** CRAT has been established under the aegis of Controller of Certifying Authorities (CCA). A CRAT consists of 1 Presiding Officer who is qualified to be a judge of a High Court or is or has been a member of the Indian Legal Service and is holding or has held a post in Grade I of that Service for at least 3 years supported by other official under him/her.

CRAT has, for the purposes of discharging its functions under the IT Act, the same powers as are vested in a civil court under the Code of Civil Procedure, 1908. However, it is not bound by the procedure laid down by the Code of Civil Procedure, 1908 but is guided by the principles of natural justice and subject to the other provisions of this Act and of any rules. CRAT has powers to regulate its own procedure including the place at which it has its sittings.

Every proceeding before a CRAT shall be deemed to be a judicial proceeding within the meaning of Sections 193 and 228, and for the purposes of Section 196 of the Indian Penal Code and the CRAT shall be deemed to be a civil court for the purposes of Section 195 and Chapter XXVI of the Code of Criminal Procedure, 1973.

**3.3.3 Centre for Development of Advanced Computing (C-DAC):** From its mission of building indigenous supercomputers in 1988, C-DAC has progressively grown and enhanced its competencies in other areas of ICTE such as multilingual computing, software technologies, electronics, RF, VLSI-ASIC and hardware design of specialized industrial products. Through in-house R&D and in synergistic collaborations with academia, research laboratories and industry, C-DAC has developed tools, technologies, products and services to meet the needs of industry and end-users in science and engineering, manufacturing and service, government, health and other strategic sectors.

Activities of C-DAC, categorised under broad thematic areas of technology, are listed below:

1. High performance computing and grid computing.
2. Multilingual and heritage computing.
3. Software technologies.
4. Professional electronics and embedded systems.
5. Cyber security and cyber forensics.
6. Health informatics.

7. Education and training.

**A. High Performance Computing (HPC) and Grid Computing:** The emphasis is on designing HPC systems using off-the-shelf components (such as CPUs) and storage along with in-house designed hardware (such as high speed networks and accelerators). C-DAC also works on system software such as light-weight communication protocols, programme development environment, and facility management tools, and a range of application software to solve problems which necessitate a massively parallel computing environment. Integration of several HPC systems with a high-bandwidth backbone network to create a national grid infrastructure is also being pursued.

**B. Multilingual and Heritage Computing:** C-DAC makes efforts towards enhancing the features of already developed tools as well as development of new tools for increased use of IT in Indian languages. These include: (a) comprehensive Indian language support for all computer applications, (b) machine-aided translation systems, (c) text-to-speech, (d) optical character and handwriting recognition, and (e) Indian language software tools to empower the physically challenged.

**C. Software Technologies:** It includes development and deployment of software tools, technologies and solutions in the areas of E-governance, web services, geomatics, multimedia computing, operating systems, and digital library.

**D. Professional Electronics, VLSI and Embedded Systems:** This area includes design of hardcore circuits, hardware, signal processing algorithms, firmware etc. for electronics products and systems. The R&D work focuses on development and validation of various re-usable intellectual property modules applying them in various electronics systems. The broad areas of power electronics, advanced wireless communications, VLSI, embedded systems, acoustics and ultrasonics, control and automation systems are also covered.

**E. Cyber Security and Cyber Forensics:** To assist in the enforcement of IT Act and to build-up expertise in tackling the menace of cyber crimes, C-DAC is a major partner of E-security Division of DIT for research and product development. Major activities include: (a) software suite for digital evidence acquisition and analysis by law enforcement authorities, (b) tools for analyzing network log files, (c) analyzing and reconstructing digital data from sessions of captured network data, (d) steganography suite to detect image files with hidden text messages, and (e) enterprise forensics and network security assurance products.

**F. Health Informatics:** C-DAC has a range of telemedicine products for open source, windows-based, web-enabled and point-to-point enterprise-specific requirements.

**G. Education and Training:** Besides technology development, C-DAC is engaged in human resources development in the area of electronics and information technology, with a range of educational programmes. The activities are aimed at enhancing the technical skills of young people at graduate level. Employment-oriented capsule programmes and diplomas in specialized software areas—such as embedded systems, geomatics, cyber security, enterprise resource management, language technology and localisation—offer value-addition to students and employed professionals to enhance their career prospects.

**3.3.4 Indian Computer Emergency Response Team (ICERT):** The mission of ICERT in is to enhance the security of India's communications and information infrastructure through proactive action and effective collaboration. Its constituency is the Indian cyber-community.

The purpose of ICERT is to become the most trusted referral agency of the Indian community for responding to computer security incidents as and when they occur. ICERT also assists members of the Indian community in implementing proactive measures to reduce the risks of computer security incidents.

**3.3.5 Centre for Materials for Electronics Technology (C-MET):** C-MET was set up as a registered scientific society in March 1990 under Department of Information Technology (formerly Department of Electronics) as a unique concept for development of viable technologies in the area of materials mainly for electronics. C-MET is operating with 3 laboratories located at Pune (headquarters), Hyderabad and Thrissur with specialized research mandate at each place.

The mission of C-MET is to develop knowledge base in the electronic materials and their processing technology and become a source of critical electronic materials, know-how and technical services for the industry and other sectors of the economy.

Core programmes of C-MET include the following:

1. Integrated electronics packaging.
2. Nano materials and devices.
3. Ultra high purity materials.
4. Materials for renewable energy.
5. Piezo sensors and actuators.

**3.3.6 Education and Research in Computer Networking (ERNET):** In this era of globalization and hyper-competition, the concept of teaching has under gone sea change. Learning and dissemination of information are becoming more important. Internet based education and E-learning are the trends of the day.

ERNET is largest nationwide terrestrial and satellite network with point of presence located at the premiere educational and research institutions in major cities of the country. Focus of ERNET is not limited to just providing connectivity, but to meet the entire needs of the educational and research institutions by hosting and providing relevant information to their users. R&D and training are integral parts of ERNET activities.

**3.3.7 Electronics and Computer Software Export Promotion Council (ESC):** ESC is India's largest electronics and IT trade facilitation organization sponsored by

Government of India. ESC has over 2,300 member companies across the length and breadth of the country in several ICT sub-sectors including computer software/service electronics hardware.

As a representative organ of Corporate India, ESC—the principal voice of Indian ICT industry both with the policy makers and the international business community—articulates the genuine, legitimate needs and interests of its member ICT exporters. ESC is India's apex institution to promote export of: (a) electronics hardware and telecom, (b) information technology, computer software and services, (c) IT-enabled services, and (d) business processing outsourcing.

Under the aegis of ESC, the IT services exports have reached to over 200 countries across the world establishing the hallmark of India's quality and competitiveness.

**3.3.8 Media Lab Asia:** Media Lab Asia is a not-for-profit company to bring the benefits of ICT to the common man. The application areas of Media Lab Asia include use of ICT for healthcare, education, livelihood and empowerment of disabled. It has a unique way of growth by working with an expanding network of leading institutions for undertaking development work. The partners are a combination of academic and R&D institutions, industry, NGOs and Government agencies. The projects can be sponsored by corporate bodies, Central and State Governments, and local municipal bodies etc. The technologies thus developed are taken from lab to land for pilot testing/deployment.

**3.3.9 National Informatics Centre (NIC):** NIC provides network backbone and E-governance support to Central and State Governments, UT Administrations, districts and other government bodies. It offers a wide range of ICT services including nationwide communication network for decentralised planning, improvement in government services and wider transparency of national and local governments. NIC assists in implementing information technology projects, in close collaboration with Central and State Governments, in the areas of: (a) Centrally-sponsored schemes and Central sector schemes, (b) State sector and State-sponsored projects, and (c) District Administration sponsored projects. NIC endeavours to ensure that the latest technology in all areas of IT is available to its users.

**3.3.10 National Informatics Centre Services Inc. (NICSI):** NICSI was set up in 1995 under National Informatics Centre (NIC) to provide total IT solutions to the Government organizations. Its main objectives are as under:

1. To promote the economic, scientific, technological, social and cultural development of India by promoting the utilization of information technology.
2. To promote further development of services, technologies, infrastructure and expertise supplementing the efforts of NIC, in directions which will increase the revenue-earning capacity of NIC.
3. To develop and promote value added computer and computer-communications services over the basic infrastructure and services developed by NIC including NICNET.

In furtherance of these objectives, NICSI has been providing following products and services to organizations in the Central and State Governments, public sector undertakings, and others:

- Hardware.
- Systems software.
- Application software.
- Software development.
- Intra-networking.
- Wide area networking.
- Video conferencing.
- Customized software.
- IT training.
- IT consultancy.
- IT implementation support.

**3.3.11 National Institute of Electronics and Information Technology (NIELIT):** Realising the wide gap in the availability of quality computer professionals to meet India's ambitious leadership plans in software exports, NIELIT has revolutionised computer knowledge dissemination, by offering its meticulously prepared curriculum through the non-formal sector—a fact that is echoed by the whole-hearted support to the curriculum by leading IT professionals and educationists. A unique feature of the curriculum is horizontal entry of students from non-accredited institutes, so that a student with a certain minimum level of computer knowledge need not start at the foundation 'O' level.

**3.3.12 National Internet Exchange of India (NIXI):** NIXI is a not-for-profit organization under Section 25 of the Companies Act 1956, and was registered on July 19, 2003. NIXI was set up for peering of ISPs among themselves for the purpose of routing the domestic traffic within the country, instead of taking it all the way to US/abroad, thereby resulting in better quality of service (reduced latency) and reduced bandwidth charges for ISPs by saving on international bandwidth. NIXI is managed and operated on a neutral basis, in line with the best practices for such initiatives globally.

**3.3.13 .in Registry (India's Official .in Domain Name Registry):** The Registry for the country code Top Level Domain (ccTLD) Name .IN is being managed by the National Internet Exchange of India (NIXI). Presently, more than 88 Registrars have been accredited to offer .IN domain name registration worldwide to customers. It has, in turn, also helped in proliferation of web hosting in the country and Indian language content in the internet. 5.5 lakh .IN Domain Names were registered till November 10, 2009. Two Data Centres have been established in Delhi and Chennai towards disaster management with an uptime of maximum 5 minutes.

**3.3.14 Society for Applied Microwave Electronics Engineering and Research (SAMEER):** SAMEER was set up as an autonomous R&D laboratory in Mumbai under the erstwhile Department of Electronics, Government of India with a broad mandate to undertake R&D work in the areas of microwave engineering and electromagnetic engineering technology. It is an offshoot of the special microwave products unit (SMPU) set up in 1977 at the Tata Institute of Fundamental Research (TIFR), Mumbai. SAMEER was set up in 1984. The centre of electromagnetics, Chennai of the erstwhile Department of Electronics was merged with SAMEER in 1987. SAMEER Kolkata centre was set up in 1994 for R&D in millimetre wave technology.

**3.3.15 Software Technology Parks of India (STPI):** STPI is a society, set up in 1991, with the objective of encouraging, promoting and boosting the software exports from India.

STPI maintains internal engineering resources to provide consulting, training and implementation services. Services cover network design, system integration, installation, operations and maintenance of application networks and facilities in varied areas ranging from VSATs to ATM-based networks.

Process development is based on the Quality Management System, which STPI also adheres to in the form of ISO 9001 certification for own process.

**3.3.16 Standardisation Testing and Quality Certification (STQC):** Established in year 1980, STQC started its services in the area of testing and calibration based on the need of small and medium sized electronic industry in the country. From a mere testing, calibration and quality assurance support to electronics hardware sector, STQC has positioned itself as a prime assurance service provider to both hardware and software sectors.

In order to provide state-of-art technology-based quality assurance services to its clients, STQC has established infrastructure and developed competent scientists to provide quality assurance and conformity assessment services in IT sector in addition to testing and calibration related services.

STQC regularly offers its value added services in area of information security, software testing/certification and development of national level assurance framework in IT and software sectors.

**3.3.17 Semiconductor Integrated Circuits Layout Design Registry (SICLDR):** SICLDR, established under the Semiconductor Integrated Circuits Layout Design Act, 2000 is the office for filing applications for semiconductor IC layout design and obtaining its registrations. Ministry of Communications and Information Technology is the administrative Ministry of the Act.

### 3.4 India as Destination for Outsourcing IT and IT-enabled Services

Starting out with basic data entry tasks, IT industry rapidly acquired a reputation as the primary low-cost destination for voice-based customer contact/support services, finance and accounting, and a range of back-office processing activities. The past few years have seen the scope of these services expanding to include increasingly more complex processes involving rule-based decision making and research/analytics services requiring informed individual judgment and domain/vertical knowledge.

Key categories within horizontal IT services include: (a) customer interaction and support (CIS), (b) finance and accounting (F&A) and other related processing services, (c) knowledge services and (d) human resource management (HRM). With experience and maturity, several providers are also beginning to develop increasing vertical specialization, thereby complementing their horizontal IT service offerings with vertical specific services.

Much of the IT work is of a routine nature such as customer contact services, but increasingly the outsourced jobs are getting sophisticated. The main knowledge services already outsourced to Indian companies are animation and simulation services, data research and analytics, and litigation services.

Other possibilities are in intellectual property research, medical content and services, pharmaceutical services, writing/content development services, database development services and end-to-end logistic invoice services.

The big impediment in India making progress from data processing and customer contact services to knowledge process services is deficiency in qualified talent in both quantitative and qualitative terms in economics and statistics, law, medical science, pharmacy, media and English language in particular. Table 3.1 shows the growth of IT-ITES professionals in India.

**Table 3.1: Growth of IT-ITES Professionals in India: Knowledge Professionals Employed***

(Number)

| | 2001-02 | 2002-03 | 2003-04 | 2004-05 | 2005-06 |
|---|---|---|---|---|---|
| IT, Engineering and R&D, software products exports | 1,70,000 | 2,05,000 | 2,96,000 | 3,90,000 | 5,13,000 |
| ITES sector | 1,06,000 | 1,80,000 | 2,16,000 | 3,16,000 | 4,09,000 |
| Domestic sector | 2,46,250 | 2,85,000 | 3,18,000 | 3,52,000 | 3,65,000 |
| Total | 5,22,250 | 6,70,000 | 8,30,000 | 10,58,000 | 12,87,000 |

*Does not include employee numbers in the hardware sector.

Source: Government of India, Planning Commission, *Eleventh Five Year Plan* (2007-12), Volume III, p. 253.

### 3.5 Strategies of Indian IT Companies

Indian companies adopt all the modes for supply of IT services abroad. The larger companies have set up wholly-owned subsidiaries in US, UK and European continent or have opened branches to facilitate the supply of services. Some of them have adopted the strategy of acquiring local companies or entering into strategic partnerships with them. They use the quality processes, innovative technologies and delivery models of acquired companies to penetrate the local market. They have permanent employees or hire consultants based in the host country.

All four modes of supply—cross-border (Mode 1), consumption abroad (Mode 2), establishment (Mode 3) and movement of natural persons (Mode 4)—are used. In the case of IT services Indian companies deploy human resources principally from India and sometimes from other foreign bases in order to provide on-site services and have also to send persons for business visits, prospecting, and sales.

Table 3.2 presents Indian IT-ITES export forecast for the period 2007-12.

**Table 3.2: Indian IT-ITES Exports Forecast 2007-12**

| Year | IT-ITES Exports (US$ billion) |
|---|---|
| 2007-08 | 37.6 |
| 2008-09 | 47.5 |
| 2009-10 | 60.0 |
| 2010-11 | 72.1 |
| 2011-12 | 86.6 |

Source: Government of India, Planning Commission, *Eleventh Five Year Plan* (2007-12), Volume III, p. 259.

In US and UK the operations are facilitated by the friendlier policies on these aspects, and the EU is known to be considering the introduction of the *blue card*–an EU-wide work permit. However, so far the immigration policies in Europe have remained problematic because of long time lines, absence of internal market mobility and non-transparent and discretionary approval processes.

A similar difference has existed between the investment procedures in US and those in the countries on the European continent. Table 3.3 shows the manpower requirements for IT-ITES in India.

**Table 3.3: Manpower Requirements for IT-ITES in India**

(Million)

| | 2002 | 2003 | 2006 | 2009 | 2012 |
|---|---|---|---|---|---|
| **IT export services** | | | | | |
| Consulting, integration, installation | 0.01 | 0.02 | 0.03 | 0.09 | 0.27 |
| IT development | 0.07 | 0.08 | 0.08 | 0.08 | 0.11 |
| Outsourced IT support | 0.09 | 0.11 | 0.17 | 0.28 | 0.53 |
| Training and education | 0.00 | 0.00 | 0.00 | 0.02 | 0.06 |
| Total | 0.17 | 0.21 | 0.29 | 0.48 | 0.97 |
| **IT-enabled services** | | | | | |
| Customer care | 0.03 | 0.05 | 0.15 | 0.42 | 0.03 |
| Finance | 0.02 | 0.03 | 0.05 | 0.09 | 0.21 |
| Human resources | 0.00 | 0.00 | 0.02 | 0.15 | 0.69 |
| Payment services | 0.00 | 0.1 | 0.05 | 0.14 | 0.45 |
| Administration | 0.2 | 0.3 | 0.5 | 0.15 | 0.15 |
| Content development | 0.03 | 0.04 | 0.07 | 0.09 | 0.20 |
| Total | 0.11 | 0.16 | 0.38 | 1.0 | 2.72 |

Source: Government of India, Planning Commission, *Eleventh Five Year Plan* (2007-12), Volume III, p. 254.

# 4

# Telecommunications in India

The telecommunication sector has witnessed a substantial growth rate since the initiation of economic reforms in 1991. Prior to this, the spread of telecommunication facilities on per capita basis was slow as compared to even other developing countries. Most of the rural areas did not have any means of telecommunication. This sector was characterised by near monopoly of the provision of services by the public sector. Even manufacturing of telecomm equipment was mainly in the hands of public sector till 1991. Since then, the manufacturing of telecom equipment has been deregulated. With the opening of the sector, the prices of some of the telecom equipments as well as provision of the services have declined. As a result, the telecom sector is likely to be one of the fastest growing sectors in the economy.

In India, telecommunication sector was the monopoly of public sector till recently. Due to high cost and poor quality of services provided by public sector, it was recognised that capacities must expand much more rapidly and competition be introduced to improve the quality of service and encourage induction of new technology. Communications has become especially important in the context of enormous growth of information technology (IT) and its impact on rest of the economy. India is perceived to have a special comparative advantage in information technology/IT-enabled services, both of which depend critically on high quality communications infrastructure.

## 4.1 Significance of Telecommunications

The telecommunications sector has undergone a total transformation throughout the world over the last two decades.

Technological advances have revolutionised the quality and range of services available. Moreover, developments in the sphere of information technology, satellite-based television broadcasting, new forms of communication such as data communication through e-mail and associated services through the internet, are transforming the way people communicate and conduct business.

Clearly, a country's ability to benefit from this revolution depends heavily on the modernity of its telecommunications network. Countries that can acquire and access information on demand and then integrate them usefully into their industrial structure through a modern telecommunications network are most likely to experience high rates of growth. Large-scale use of information and telecommunications technologies directly influences productivity, cost effectiveness and competitiveness in industries with high levels of product differentiation and low levels of unit prices. In these industries, prompt availability of information about demand trends or price movements can boost competitive advantage. International competitiveness of the textile, garment, toy and consumer electronics industries in Taiwan and the Philippines have been found to be effective due to advanced telecommunications services that allow tight links between commercial distribution in American and European markets and local production.

An advanced telecommunications system is equally important for service industries like banking, trading, retailing, transportation, maintenance and insurance where information and real-time communication are vital to the production process. A reduction in the costs of these services will indirectly enhance international competitiveness within the entire economic system, since lower marketing costs mean lower costs for manufacturing firms exposed to international trade. The extraordinary efficiency of financial markets of Hong Kong and Singapore is based on the extensive use of such advanced telecommunications services.

Thus, improved communication is vital for productivity in

all spheres of activity: agriculture, industry, trade and commerce. In addition, with the increasing movement of people within and outside the country, easy communication is essential for enhancing human welfare so that families, friends and acquaintances can overcome the barriers of distance. For a developing economy like India, it is thus important that policies and programmes be initiated, and resources committed to bring about rapid growth in the sector.

### 4.2 Department of Telecommunications (DoT)

The telecom sector in India has been witnessing a continuous process of reforms since 1991. With the opening of international long distance services and internet telephony from April 2002, the process of liberalization and opening up the sector for competition was completed. Apart from opening up the basic as well as value-added market segment of telecom services to the private sector, the Government has taken major initiatives in offering a level playing field to operators by constituting the Telecom Regulatory Authority of India (TRAI). However, a great deal needs to be done, especially in terms of the implementation strategy so that the country can enjoy the benefits of the reforms.

Driven by various policy initiatives, which included National Telecom Policy (NTP) 1994, NTP, 1999, NTP, 2011, and NTP, 2012, the Indian telecom sector has witnessed a complete transformation in recent years. During this period, it has achieved a phenomenal growth and is poised to take a quantum jump. For bringing further accelerated growth along with modernization in line with the trends in other developed countries the telecommunication sector demands initiation of further policy measures and induction of state-of-the-art technologies. The most striking success of liberalization policy is visible in telecommunications. Teledensity has risen dramatically, an entire industry with private players has come about, privatisation has begun with VSNL, and new technologies have come about. Looking forward, the telecom sector will feature lively competition between multiple private firms, with a strong role for TRAI to establish pro-competitive policies.

DoT is a Department of the Ministry of Communications and Information Technology, Government of India. It is entrusted with the following functions:

1. Policy, licensing and co-ordination matters relating to telegraphs, telephones, wireless, data, facsimile and telematic services and other like forms of communications.
2. International co-operation in matters connected with telecommunications including matters relating to all international bodies dealing with telecommunications such as International Telecommunication Union (ITU), its Radio Regulation Board (RRB), Radio Communication Sector (ITU-R), Telecommunication Standardization Sector (ITU-T), Development Sector (ITU-D), International Telecommunication Satellite Organization (INTELSAT), International Mobile Satellite Organization (INMARSAT), Asia Pacific Telecommunication (APT).
3. Promotion of standardization, research and development in telecommunications.
4. Promotion of private investment in telecommunications.
5. Financial assistance for the furtherance of research and study in telecommunications technology and for building up adequately trained manpower for telecom programme, including: (a) assistance to institutions, assistance to scientific institutions and to universities for advanced scientific study and research, and (b) grant of scholarships to students in educational institutions and other forms of financial aid to individuals including those going abroad for studies in the field of telecommunications.
6. Procurement of stores and equipment required by the Department of Telecommunications.
7. Telecom Commission.
8. Telecom Regulatory Authority of India.
9. Telecom Disputes Settlement and Appellate Tribunal.

10. Administration of laws with respect to any of the matters specified in this list, namely: (a) Indian Telegraph Act, 1885, (b) Indian Wireless Telegraphy Act, 1933, and (c) Telecom Regulatory Authority of India Act, 1997.
11. Indian Telephone Industries Limited.
12. Post-disinvestment matters relating to M/s Hindustan Teleprinters Limited.
13. Bharat Sanchar Nigam Limited.
14. Mahanagar Telephone Nigam Limited.
15. Videsh Sanchar Nigam Limited and Telecommunications Consultants (India) Limited.
16. All matters relating to Centre for Development of Telematics (C-DOT).
17. Residual work relating to the erstwhile Department of Telecom Services and Department of Telecom Operations, including matters relating to: (a) cadre control functions of Group A and other categories of personnel till their absorption in Bharat Sanchar Nigam Limited, and (b) administration and payment of terminal benefits.
18. Execution of works, purchase and acquisition of land debitable to the capital Budget pertaining to telecommunications.

### 4.3 Organizations under DoT

**4.3.1 Public Sector Units:** These are as under:

**A. Bharat Sanchar Nigam Limited (BSNL):** BSNL provides a cost-effective alternative to mobile subscribers with low entry charges, no security deposit for local/ISD connection, no CLIP charges, no additional airtime charges while roaming in BSNL network, free zonal roaming for pre-paid subscribers, no monthly charges for national roaming under Plan 325, free SMS services for both incoming and outgoing to all prepaid customers and additional grace-period of 90 days for SIM-activation without activation charges.

In order to supplement the efforts of public sector providers and to ensure greater competition in providing basic telephone services, private companies are being licensed to offer basic services.

**B. Indian Telephone Industries Limited (ITI):** India's first Public Sector Unit (PSU)-ITI Ltd was established in 1948. Ever since, as a pioneering venture in the field of telecommunications, it has contributed to 50 percent of the present national telecom network. With state-of-the-art manufacturing facilities spread across six locations and a countrywide network of marketing/service outlets, the company offers a complete range of telecom products and total solutions covering the whole spectrum of Switching, Transmission, Access and Subscriber Premises equipment.

ITI joined the league of world class vendors of Global System for Mobile (GSM) technology with the inauguration of mobile equipment manufacturing facilities at its Mankapur and Rae Bareli Plants in 2005-06. This ushered in a new era of indigenous mobile equipment production in the country. These two facilities supply more than nine million lines per annum to both domestic as well as export markets.

The company is consolidating its diversification into Information and Communication Technology (ICT) to hone its competitive edge in the convergence market by deploying its rich telecom expertise and vast infrastructure. Network Management Systems, Encryption and Networking Solutions for Internet Connectivity are some of the major initiatives taken by the company.

Secure communications is the company's forte with a proven record of engineering strategic communication networks for India's Defence forces. Extensive in-house R&D work is devoted towards specialized areas of Encryption, NMS, IT and Access products to provide complete customized solutions to various customers.

**C. Mahanagar Telephone Nigam Limited (MTNL):** MTNL was set up on April 1, 1986 by the Government of India to upgrade the quality of telecom services, expand the telecom network, introduce new services and to raise revenue for telecom development needs of India's key metros, viz. Delhi, the political capital and Mumbai, the business capital. The company has taken rapid strides to emerge as India's leading and one of Asia's largest telecom operating companies. The company has also been in the forefront of technology induction by converting 100 percent of its telephone exchange network into the state-of-the-art digital mode. The Government of India currently holds 56.25 percent stake in the company.

**D. Telecommunications Consultants India Limited (TCIL):** TCIL, a prime engineering and consultancy company, is a wholly owned Government of India Public Sector Enterprise under the administrative control of the Department of Telecommunications (DOT), Ministry of Communications and Information Technology, Government of India. TCIL was set up in 1978 for providing Indian telecom expertise in all fields of telecom, civil and IT to developing countries around the world. Company's core competence is in the fields of switching, transmission systems, cellular services, rural telecommunication, optical fibre based backbone transmission systems, IT and networking solutions, application software, e-governance, 3G network, WIMAX technology and also civil construction projects.

**4.3.2 Other Organizations:** These are as under:

**A. Wireless Planning and Coordination Wing (WPC):** The WPC Wing of the Ministry of Communications, created in 1952, is the National Radio Regulatory Authority responsible for Frequency Spectrum Management, including licensing and caters for the needs of all wireless users (Government and Private) in the country. It exercises the statutory functions of the Central Government and issues licenses to establish, maintain and operate wireless stations. WPC is divided into major sections like licensing and regulation (LR), new technology group (NTG) and Standing Advisory Committee on Radio Frequency Allocation (SACFA). SACFA makes the recommendations on major frequency allocation issues, formulation of the frequency allocation plan, making recommendations on the various issues related to International Telecom Union (ITU), to sort out problems referred to the committee by various wireless users, Siting clearance of all wireless installations in the country etc.

**B. Telecom Engineering Centre (TEC):** It is a technical body representing the interest of Department of Telecommunications, Government of India. It performs the following functions:

1. Specification of common standards with regard to telecom network equipment, services and interoperability.
2. Generic requirements (GRs) and interface requirements (IRs).
3. Issuing interface approvals, certificate of approvals, service approvals and type approvals.
4. Formulation of standards and fundamental technical plans.
5. Interact with multilateral agencies ITU for standardisation.
6. Develop expertise to imbibe the latest technologies and results of R&D.
7. Provide technical support to DoT and technical advice to TRAI.
8. Co-ordinate with C-DOT on the technological developments in the telecom sector for policy planning by DOT.

**C. Telecom Enforcement, Resource and Monitoring (TERM) Cells (Formerly Known as Vigilance Telecom Monitoring [VTM] Cells):** With the increasing number of telecom operators in the country, the Government felt the need for presence of Telegraph Authority in the field at all the Licence Service Areas and Large Telecom Districts of the country, in order to ensure that service providers adhere to the licence conditions and for taking care of telecom network security issues. With the growth of private telecom and internet services, an increase in illegal/clandestine telecom operations was also observed. To address these issues, the Government created initially 4 Vigilance Telecom Monitoring (VTM) Cells in November 2004 in Delhi, Mumbai, Hyderabad and Chennai. 9 more VTM Cells were created during 2006 for the circles of Punjab, Rajasthan, Gujarat, Kerala, Karnataka, Maharashtra, Tamil Nadu, West Bengal and Uttar Pradesh (East) and 15 VTM Cells were subsequently added in January 2007 for Andhra Pradesh, Bihar, Madhya Pradesh, Haryana, Uttar Pradesh (West), Andaman and Nicobar, Assam, Chhattisgarh, Jammu and Kashmir, Jharkhand, Himachal Pradesh, North East-I, North East-II, Orissa and Uttarakhand. 6 more VTM Cells were added in March 2007 for Kolkata, Ahmedabad, Bangalore, Pune, Jaipur and Lucknow, taking the total number of VTM Cells to 34.

Since the formation of VTM Cells in the DoT, the role and functions of VTM Cells have increased many-fold. With a view to reflect the entire gamut of functions assigned to VTM Cells and to distinguish their role vis-à-vis staff-vigilance activities, the name of VTM Cells was changed to Telecom Enforcement, Resource and Monitoring (TERM) Cells w.e.f August 5, 2008.

**D. Centre for Development of Telematics (C-DoT):** It is the telecom technology development centre of the Government of India. It was established in August 1984 as an autonomous body. It was vested with full authority and total flexibility to develop state-of-the-art telecommunication technology to meet the needs of the Indian telecommunication network.

Following are the objectives of C-DoT:

1. Work on telecom technology products and services.
2. Provide solutions for current and future requirements of telecommunication and converged networks including those of national importance, especially related to rural applications, strategic sector and security agencies etc.
3. Provide market orientation to R&D activities and sustain itself as a centre of excellence.
4. Build partnerships and joint alliances with academia, industry, solution providers, and other R&D organizations to offer cost-effective solutions.
5. Support service providers in the introduction of new

technologies, features and services by optimal utilization of installed networks, pilots and studies.

6. Strengthen the Indian telecom manufacturing base, by transfer of technologies developed by it.

## 4.4 National Telecom Policy (NTP), 1999

The NTP announced in 1999 provided the basic framework for the future development and growth of the telecom sector in the country. The NTP took into account the far-reaching technological developments taking place in the telecom sector globally and stressed on implementing the Governments' resolve to make India a global IT superpower. The objectives of the NTP were the following:

1. To make available affordable and effective communications for the citizens.
2. To strive to provide a balance between the provision of universal service to all uncovered areas, including the rural areas and the provision of high-level services capable of meeting the needs of the country's economy.
3. To encourage the development of telecommunication facilities in remote, hilly and tribal areas of the country.
4. To create a modern and efficient telecommunications infrastructure taking into account the convergence of IT, media, telecom and consumer electronics and thereby propel India into becoming an IT superpower.
5. To convert public call offices (PCOs), wherever justified, into public teleinfo centres offering multimedia services like integrated services digital network (ISDN) services, remote database access, government and community information system etc.
6. To transform in a time bound manner, the telecommunications sector to a greater competitive environment in both urban and rural areas providing equal opportunities and level playing field for all players.
7. To strengthen research and development (R&D) efforts in the country and provide an impetus to build world-class

manufacturing capabilities.

8. To achieve efficiency and transparency in spectrum management.
9. To protect the defence and security interests of the country.
10. To enable Indian telecom companies to become truly global players.

NTP, 1999 envisaged that India should emerge as a major manufacturing base and major exporter of telecom equipment. The Government no longer engages in telecom service provision which is now the task of private and public sector companies operating in a competitive market. The Telecom Regulatory Authority of India (TRAI) now performs the regulatory functions of the sector. The Department of Telecom and the Telecom Commission have re-focused on the functions of policy formulation, licensing, wireless spectrum management, administrative monitoring of PSUs, research and development and standardization/validation of equipment etc.

**4.5 National Telecom Policy (NTP), 2011 and 2012**

Telecommunication has emerged as a key driver of economic and social development and in an increasingly knowledge intensive global scenario India needs to play a leadership role. NTP, 2011 is designed to ensure that India plays this role effectively and transforms the socio-economic scenario through accelerated equitable and inclusive economic growth by laying special emphasis on providing affordable and quality telecommunication services in rural and remote areas.

The thrust of this policy is to underscore the imperative that sustained adoption of technology would offer viable options in overcoming developmental challenges in education, health, employment generation, financial inclusion and much else. NTP, 2011 is an initiative to create a conducive policy framework to address these issues and to touch lives of all citizens and transform India. By formulating a clear policy regime, NTP, 2011 endeavours to create an investor friendly environment for attracting additional investments in the sector apart from generating manifold employment opportunities in various segments of the sector. In achieving the goal of NTP, 2011 revenue generation will play a secondary role.

NTP, 2011 has the vision of *Broadband on Demand* and envisages leveraging telecom infrastructure to enable all citizens and businesses, both in rural and urban landscape, to participate in the Internet and web economy thereby ensuring equitable and inclusive development across the nation.

NTP, 2011 provides a roadmap for India to become a leader in cutting edge, state-of-the art technologies through R&D and creation and incorporation of Indian IPRs in global standards.

NTP, 2011 envisions providing the people of India, secure, reliable, affordable and high quality converged telecommunication services anytime, anywhere.

In short, the primary objective of NTP, 2011 is maximizing public good by making available affordable, reliable and secure telecommunication and broadband services across the entire country. The main thrust of the Policy is on the multiplier effect and transformational impact of such services on the overall economy. It recognizes the role of such services in furthering the national development agenda while enhancing equity and inclusiveness. Direct revenue generation would continue to remain a secondary objective. NTP, 2011 also recognizes the predominant role of the private sector in this field and the consequent policy imperative of ensuring continued viability of service providers in a competitive environment. Pursuant to NTP, 2011, these principles would guide decisions needed to strike a balance between the interests of users/consumers, service providers and government revenue.

The National Telecom Policy 2012 (NTP, 2012) has been conceived with a vision to transform the country into an empowered and inclusive knowledge-based society, using telecommunications as a platform.

NTP 2012 envisages the broadband facilities to be provided both on fixed and wireless networks. A National Optic Fibre Network will be established, connecting all the Gram Panchayats in the first phase and all the villages/habitations in the second phase, both of which will be completed in a definite time frame. The policy also seeks to make available adequate spectrum for wireless broadband services.

### 4.6 Telecommunications Regulatory Authority of India (TRAI)

TRAI performs the regulatory functions of telecom sector. It has taken various measures in recent years with a view to obtain a sound framework for competition between multiple technologies and vendors:

1. TRAI has given two recommendations on VSAT. First relates to issuance of licenses for VSAT services such as 'Receive only VSAT' because it has great potential to open new vistas of innovation in information sharing and spread of education and in facilitating the efforts of the government in bridging the Digital Divide. The second recommendation relates to facilitating VSAT operations by permitting Higher Data Rate, Reduction in Licence fee for Captive VSAT Network and reducing minimum antennae size. In order to promote internet in the country, TRAI has made a number of recommendations including setting up of National Internet Exchange of India (NIXI). In order to ensure for the consumers equal ease of access to different domestic and international long distance service providers, TRAI has issued a directive to all access providers and national/international long distance operators regarding implementation of carrier selection in their respective network. A regulatory principle was maintained to prevent any anti-competitive or predatory behaviour on part of dominant operators.
2. The quality of services of all service providers is being monitored by TRAI on the basis of quarterly reports

submitted by them. TRAI is also conducting an objective survey of quality of service (QoS) parameters of basic and cellular service in different circles through an independent agency. The agency has also been assigned the task of making an assessment of customer satisfaction through a survey. TRAI has also issued a regulation on quality of service for voice over internet protocol (VOIP) based international long distance service.

3. It has recommended deregulating regulatory intervention in cellular tariffs. This implies that operators no longer need to have prior approval of the regulator for implementing tariff plans except under certain conditions. Guidelines were also issued to service providers in respect of pre-paid cards which would enable increased consumer awareness and transparency for the subscriber.
4. It has finalized the system of accounting separation (SITUATION), providing a detailed accounting and financial system maintained by telecom service providers and has forwarded the same to DoT for notification under Section 35 (2) (d) of the TRAI Act. The proposed system, based on the principles of accounting separation, would help monitor and measure financial performance of individual telecom products/network services and help disinvestment aggregate costs to the level of network elements. It would also help in identification of cross-subsidisation practices in the industry, wherever these exist. By providing for maintenance of detailed cost records right up to the level of network elements, the system will help generate accurate information on costs, which is necessary for tariff and interconnected regulations. TRAI's document on accounting separation would go a long way in promoting healthy competition with adequate mechanisms built in to detect unfair competitive practices. While the system is mainly aimed at enabling better monitoring and control by TRAI, the new system is also expected to help operators establish a more

comprehensive system for arriving at product/network service wise costs and revenues and thus enhance their operating efficiency.

5. The National and International Long Distance markets in telecom have been opened up to competition. These policy measures have resulted in significant reduction in long distance tariffs due to competitive pressures. This has drastically reduced the margin available to fund the Access Deficit incurred. In a multiple operator environment, it is important to specify an International Usage Charge (IUC) regime which gives greater certainty to the Inter-operator settlements and facilitates inter-connection agreements. Therefore, TRAI notified its Telecommunication Inter-connection Usage Charges (IUC) Regulation, 2003 in January 2003 which covered arrangements among service providers for payment of inter-connection usage charges for telecommunication services, covering basic services which include WLL (M) service, cellular mobile service providers and long distance operators throughout the territory of India. This regulation has specified the IUC for origination, transit and termination in a multi-operator environment.

### 4.7 Agenda for Reforms

Telecommunications is one of the fastest growing sectors in India. However, viewed in the context of global growth patterns and indicators, the sector is still in the early stages of development. The sector needs, especially in terms of broad-band, to expand at substantially higher rates to meet the needs of related sectors like information technology (IT), information and broadcasting (I&B) and other sectors of the economy. Keeping this perspective in view, the sector needs to be treated essentially as an infrastructure sector. Once the required teledensity is achieved and the necessary support network has been created, the sector could be treated as service sector.

Telecommunications should continue to be treated as a

major element of infrastructure for growth and development of the Indian economy. For rapid and balanced growth of the sector, the Government should not view it as an opportunity area for resource mobilisation. This needs to be reflected in its licensing and taxation policies.

Inter-circle long-distance services should be opened to the private sector. The process leading to this action should incorporate appropriate lessons from the experience of privatization of the telecom services in India and abroad.

As the Indian telecom network is already one of the largest in the world, it is important that the country starts investing in developing equipment manufacturing capabilities. This will help India establish itself as a manufacturing base that can take care of the future telecom requirements of the country as well as export sales. Rationalisation of import tariff should be extended to components and inputs required by manufacturers of such products so as to encourage domestic production. India should emerge as a major manufacturing base and an exporter of telecom equipment.

A major development in the recent past has been the convergence of telecom, broadcasting and information technology. Primarily driven by emerging technologies, the information era is fast demolishing the traditional walls between telecom, broadcasting and computer technology in the marketplace. Government policymakers have to take cognisance of this happening and shape future policies accordingly.

With a view to ensuring optimum growth in the coming years, Government's broad policy of taxes and regulation for the telecom sector has to be a promotional one. Mopping up of resources or revenue generation by the Government should not be a determinant of the policy governing the sector. The incidence of licence fees in the form of revenue sharing and spectrum charges has to be guided by this principle. Keeping in line with the policy adopted by most of the progressive administrations in the world, the licence fee needs to be aligned with the cost of regulation and administration of Universal Service Obligations (USO). As part of the promotional policy, there is need for the TRAI to work out afresh, the revenue shares and USO regime.

Private investment is also expected to play a leading role in the expansion of telecom services. In the area of value added services, the private sector would continue to play a dominant role. The quantum of investment by the private operators would basically get determined by the rate of return on such investments—both basic as well as value added services. Foreign direct investment (FDI) also has a major role to play in supplementing the resources of the domestic private sector as the scale of investment envisaged is large. To boost private sector investment, appropriate policy initiatives should to be undertaken.

India is a preferred location for software vendors for its quality and cost. India has strong Unix base which provides opportunity for the development of products for internet-based applications. Further, India has global connectivity with international dialling facility from over 13220 locations, leased/switched high-speed data links from major centres through STPs and VSNL. Internet connectivity is provided through several networks. Abundant investment opportunities exist in India in the following thrust areas:

1. Communication infrastructure.
2. Optic fibre cable.
3. Gateways.
4. Satellite-based communication wireless.
5. Software development.
6. IT-enabled services.
7. IT-education.
8. IT-enabled education.
9. Data Centres and server farms.

The future policies and programmes for the telecom sector should be guided by the basic goal of creating a world-class telecom infrastructure in order to meet the requirements of IT-based sector and needs of a modernizing economy on the least cost basis. Ensuring value for money to the consumers and easy and affordable access to basic telecom services to every one and every where, should be the other goal of policies. The major objectives for the future should include the following:

1. Affordable and effective communication facilities to all citizens.
2. Provision of universal service to all uncovered areas, including rural areas.
3. Building a modern and efficient telecommunications infrastructure to meet the convergence of telecom, IT and the media.
4. Transformation of the telecommunications sector to a greater competitive environment, providing equal opportunities and level playing field for all the players.
5. Strengthening R&D efforts in the country.
6. Achieving efficiency and transparency in spectrum management.
7. Protecting the defence and security interests of the country.
8. Enabling Indian telecom companies to become truly global players.

It is now imperative to move towards convergence between telecom, broadcast and IT services, networks, platforms, technologies and overcome the existing segregation of licensing, registration and regulatory mechanisms in these areas to enhance affordability, increase access, delivery of multiple services and reduce cost.

The emerging technology trends in electronics hardware, telecom connectivity and IT will make it possible for millions of citizens to access services electronically in self-service mode using mobile phones and the Internet or through assisted service points such as common service centres etc.

A concerted effort to boost manufacturing activity is now exigent as robust economic growth in the country is leading to an extraordinarily high demand for electronic products in general and telecom products in particular.

To sum up, telecommunications is one of the prime support services needed for rapid growth and modernization of various sectors of the economy. It has become especially important in recent years because of enormous growth of information technology (IT) and its significant impact on the rest of the economy. India is perceived to have a special comparative advantage in IT and IT-enabled services. However, sustaining this advantage depends critically on high quality telecommunication infrastructure. Keeping this in view, the focus has to be on the provision of world class telecommunication facilities at reasonable rates.

Provision of telecom services in rural areas would be another thrust area to attain the goal of accelerated economic development and social change. Although the telecom network has grown rapidly in recent years, its growth needs to be accelerated further. It is equally important to speed up the structural changes in this sector in line with trends in other countries to ensure that telecommunication services are not only made available on the scale needed to sustain rapid growth in the economy as a whole but also that their cost are in tune with the expectations of a modernizing economy.

India's vast size, the large number of spread out settlements and its large unserved population provides a huge potential for the expansion of telecommunication services. This opportunity must be seized by: (a) raising substantially the penetration ratio in order to provide access to dependable voice communication means with at least national subscriber trunk dialling (STD) connectivity to a much larger cross-section of the population, (b) satisfying the more demanding audio, video and data communication needs of the business community in all major business districts in the country, and (c) making provision for easy up-gradation of the network to meet future communication needs.

All this would need a rapid expansion and upgradation of the existing network. India has also expressed its first commitment to make large investments in value-added services by opening up this sector. All this would place India among the leading countries in terms of equipment purchase. Concerted efforts need to be made to attract foreign companies to set up manufacturing bases in India for supporting domestic as well as export sales.

Substantial investments will be required to ensure that India acquires the status of a global player in telecom. This cannot be achieved through governmental efforts or through a monopoly state-owned operator alone. The Government of India has realised the ground realities and initiated the process of progressive deregulation of the telecom sector.

The manufacturing capacity of the indigenous industry is small in relation to the other major operators in the world and exports constitute a very small proportion of the total production of telecom equipment in the country. Promoting exports as a thrust area and development of Indian multinationals should be among the major goals in this sector. To achieve this objective, Government should aim at removing the various constraints relating to transfer of the latest technology, access to cheap international finance, joint ventures with foreign companies, rationalization of custom and import duties on inputs and development of a strong industry-sponsored R&D base. Due to fast changing technology in this sector and the existing technological gap in relation to the developed countries, India will have to rely on transfer of technology in a big way in the immediate future. To achieve this, strategic alliance with leading international companies will have to be an integral part of the overall exercise of technology transfer.

# 5

# Technology Standards for Internet

## 5.1 Basic Structure and Topology of Internet

Internet is a vast and connected network of individual computers and computer networks for the purpose of communicating with each other, using the same communication protocol—Transmission Control Protocol/Internet Protocol (TCP/IP). When two or more computers are connected a network is created. When two or more networks are connected, an *inter-network* (or Internet) is created. The Internet, as commonly understood, is the largest example of such a system. Internet is often and aptly described as *Information Superhighway*, a means to reach innumerable potential destinations. The destination can be any one of the connected networks and host computers.

## 5.2 Origin of Internet

Internet has evolved to its present state out of a US Department of Defence project named Advanced Research Project Administration Network (ARPANet), developed in the late 1960s and early 1970s as an experiment in wide area networking. A major perceived advantage of ARPANet was that the network would continue to operate even if a segment of it is lost or destroyed since its operation did not depend on operation of any single computer. Though originally designed as a defence network, over the years it was used predominantly in areas of scientific research and communication. By the 1980s, it moved out of Pentagon's control and more independent networks from US and outside got connected to it.

In 1986, the US National Science Foundation (NSF) established a national network based on ARPA protocol using commercial telephone lines for connectivity. The NSFNet was accessible by a much larger scientific community, commercial networks and general users and the number of host computers grew rapidly. Eventually, NSFNet became the framework of present day Internet. ARPANet was officially decommissioned in 1990.

It has become possible for innumerable computers operating on different platforms to communicate with each other over Internet because they adopt the same communication protocol, viz. TCP/IP. It is a set of rules which define how computers communicate with each other. In order to access Internet one must have an account in a host computer, set up by any one of the Internet Service Providers (ISPs). The accounts can be Serial Line Internet Protocol (SLIP) or Point to Point Protocol (PPP) account. These accounts allow creating temporary TCP/IP sessions with the host, thereby allowing the computer to join the Internet and directly establish communication with any other computer in the Internet. Through this type of connection, the client computer does not merely act as a remote terminal of the host, but can run whatever programmes are available on the web. It can also run several programmes simultaneously, subject to limitations of speed and memory of the client computer and modem. TCP/IP protocol uses a unique addressing scheme through which each computer on the network is identified.

TCP/IP protocol is insecure because data packets flowing through TCP/IP networks are not normally encrypted. Thus, any one who interrupts communication between two machines will have a clear view of the data, passwords and the like. This has been addressed through Secured Socket Layer (SSL), a Transport Layer Security (TLS) system which involves an encrypted session between the client browser and the web server.

## 5.3 Major Uses of Internet

These are the following:

**5.3.1 File Transfer Protocol (FTP):** It is a mechanism for transferring files between computers on the Internet. It is possible to transfer a file to and from a computer (FTP site) without having an account in that machine. Any organization intending to make available to public its documents would normally set up a FTP site from which any one can access the documents for download. Certain FTP sites are available to validated users with an account ID and password.

**5.3.2 Electronic-mail (E-mail):** The most common and basic use of Internet is the exchange of e-mail. It is an extremely powerful and revolutionary result of Internet which has facilitated almost instantaneous communication with people in any part of the world. With enhancements like attachment of documents, audio, video and voice mail, this segment of Internet is fast expanding as the most used communication medium for the whole world. Many websites offer e-mail as a free facility to individuals. Many corporates have interfaced their private networks with Internet in order to make their e-mail accessible from outside their corporate network.

**5.3.3 World Wide Web (WWW):** Internet encompasses any electronic communication between computers using TCP/IP protocol, such as e-mail, file transfers etc. WWW is a segment of Internet, which uses hyper text mark-up language (HTML) to link together files containing text, rich text, sound, graphics, video etc. and offers a very convenient means of navigating through the net. It uses hypertext transfer protocol (HTTP) for communication between computers. Web documents, which are referred to as pages, can contain links to other related documents in a tree like structure. The person browsing one document can access any other linked page. The web documents and the web browsers which are the application programmes to access them, are designed to be platform independent. Thus, any web document can be accessed irrespective of the platform of the computer accessing the document and that of the host computer. The *point and click* method of browsing is extremely simple for any lay user of the net. In fact, the introduction of web since early 1990 has made Internet an extremely popular medium and its use in business has been enhanced dramatically.

The next in the HTML genre is the extensible mark-up language (XML), which allows automated two-way information flow between data stores and browser screens. XML documents provide both the raw content of data and the data structure and are projected by its proponents as taking the web technology beyond the limits of HTML.

**5.3.4 Wireless Application Protocol (WAP):** WAP is the latest industry standard which provides wireless access to Internet through hand-held devices like a cellular telephone. This is an open standard promoted by WAP forum and has been adopted by all major handset manufacturers of the world. WAP is supplemented by wireless application environment (WAE), which provides industry-wise standard for developing applications and services for wireless communication networks. This is based on WWW technology and provides for application for small screens, with interactive capabilities and adequate security. Wireless transaction protocol (WTP), which is the equivalent of TCP, sets the communication rules and wireless transport layer security (WTLS) provides the required security by encrypting all the session data. WAP is set to revolutionize the commercial use of net.

## 5.4 Security Concerns

One of the biggest attractions of Internet, as an electronic medium, is its openness and freedom. It is a public domain and there is no restriction on who can use it as long as one adheres to its technical parameters. This has also given rise to concerns over the security of data and information and privacy. These concerns are common to any network including closed user group networks. However, over the Internet, the dimensions of risk are larger while the control measures are relatively fewer. The key components of such concern are as under:

1. Authentication, i.e. assurance of identity of the person in a

deal.

2. Authorization, i.e. a party doing a transaction is authorized to do so.
3. Privacy or confidentiality of data, information etc. relating to any deal.
4. Data integrity, i.e. assurance that the data has not been altered.
5. Non-repudiation, i.e. a party to the deal cannot deny that it originated the communication or data.

## 5.5 Technology for Internet Banking

The Internet has provided a new and inexpensive channel for banks to reach out to their customers. It allows customers to access facilities provided by banks round the clock and 7 days a week. It also allows customers to access these facilities from remote sites/home etc. However, all these capabilities come with a price. The highly unregulated Internet provides a less than secure environment for the banks to interface. The diversity in computer, communication and software technologies used by the banks vastly increases the challenges facing the online bankers. The banks planning to offer Internet-banking should have explicit policies on security.

**5.5.1 Computer Networking and Internet:** The purpose of computer networking is sharing of computing resources and data across the whole organization and the outside world. Computer networks can be primarily divided into the following two categories based on speed of data transfers and geographical reach:

1. A local area network (LAN) connects many servers and workstations within a small geographical area, such as a floor or a building.
2. A wide area network (WAN), on the other hand, is designed to carry data over great distances and is generally point-to-point. Connectivity in WAN set up is provided by using dial-up modems on the public switched telephone network (PSTN), very small aperture terminal (VSAT)

networks, integrated services digital network (ISDN) or by using virtual private networks (VPN) which are software-defined, dedicated and customized services used to carry traffic over the Internet.

**5.5.2 Protocols:** The data transmission protocol suite used for the Internet is known as transmission control protocol/internet protocol (TCP/IP). The Internet is primarily a network of networks. The networks in a particular geographical area are connected into a large regional network. The regional networks are connected via a high speed *backbone*. The data sent from one region to another is first transmitted to a Network Access Point (NAP) and then routed over the backbone. Each computer connected to the Internet is given a unique IP address and a hierarchical domain name.

The Internet can be accessed using various application-level protocols such as file transfer protocol (FTP), remote terminal control protocol (Telnet), simple mail transport protocol (SMTP), hypertext transfer protocol (HTTP). These protocols run on top of TCP/IP. The most innovative part of the Internet is the world wide web (WWW). The web uses hyperlinks, which allow users to move from any place on the web to any other place. The web consists of web pages, which are multimedia pages composed of text, graphics, sound and video. The web pages are made using hypertext mark-up language (HTML). The web works on a client-server model in which the client software, known as the browser, runs on the local machine and the server software, called the web server, runs on a possibly remote machine. Some of the popular browsers are Microsoft Internet Explorer and Netscape Navigator. With the popularity of web, organizations find it beneficial to provide access to their services through the Internet to its employees and the public.

**5.5.3 Application Architecture:** A computer-based application may be built as a monolithic software, or may be structured to run on a client–server environment, or even have three or multi-tiered architecture. A computer application typically separates its following 3 main tasks:

1. Interactions with the user.
2. Processing of transactions as per the business rules.
3. Storage of business data.

The three tasks can be viewed as three layers, which may run on the same system (possibly a large, proprietary computer system), or may be separated on to multiple computers (across the Internet), leading to three-tier or multi-tier architecture. These layers can be briefly described as follows.

**A. Presentation Layer:** This layer is responsible for managing the front-end devices, which include browsers on personal computers, personal digital assistants (PDAs), mobile phones, internet kiosks, web TV etc.

The presentation layer takes care of user interface-related issues like display details, colour, layout, image etc. It also has important responsibilities in user authentication and session management activity.

**B. Application Layer:** It contains the business logic (for processing of data and transactions) and necessary interfaces to the data layer. It processes requests from the presentation layer, connects to the data layer, receives and processes the information and passes results back to the presentation layer. It is responsible for ensuring that all the business rules are incorporated in the software. The issues of scalability, reliability and performance of the services to a great extent depend upon the application layer architecture.

**C. Data Layer:** The data layer uses a database package to store, retrieve and update application data. The database may be maintained on one or multiple servers. A database package also supports back-up and recovery of data, as well as logging of all transactions.

**5.5.4 Precautions in the Administration of Systems and Applications:** The role of the network and the database administrator is pivotal in securing the information systems of any organization. The role extends across various job functions and any laxity in any of the functions leaves the system open for malicious purposes. A few important functions of the administrator and how they relate to or impinge on system security are discussed below.

**A. Installation of Software:** A software (whether system or application) needs to be carefully installed as per the developer's instructions. The software system may contain bugs and security holes, which over a period are fixed through appropriate patches. It is necessary to know the latest and correct configuration of all software packages. Hackers and intruders are often aware of these bugs and may exploit known weaknesses in the software. Hence, care should be taken to install only the latest versions of software with the latest patches.

Further, improper installation may lead to degradation of services. Installation of pirated software is not only illegal and unethical, but may also contain Trojans and viruses which may compromise system security. In the case of installation of outsourced software, care should be taken to compare the source code and the executable code using appropriate tools as unscrupulous developers may leave backdoor traps in the software for illegal access and update to the data. Besides, while installing software care should be taken that only necessary services are enabled on a need-to-use basis.

**B. Access Controls and User Maintenance:** An administrator has to create user accounts on different computer systems and give various access permissions to the users. Setting access controls to files, objects and devices reduces intentional and unintentional security breaches. System policy of a bank should specify access privileges and controls for the information stored on the computers. The execution privilege of most system-related utilities should be limited to system administrators so that users may be prevented from making system level changes. The write/modify access permissions for all executables and binary files should be disabled. If possible, all log files should be made *append only*. All sensitive data should be made more secure by using encryption. The system and database administrators are also responsible for the maintenance of users and the deletion of inactive users. Proper logs should be maintained of dates of user creation and validity period of users. There should be a frequent review to identify unnecessary users and privileges, especially of temporary users such as system maintenance personnel and system auditors.

**C. Backup, Recovery and Business Continuity:** Backup of data, documentation and software is an important function of the administrators. Both data and software should be backed up periodically. The frequency of back up should depend on the recovery needs of the application. Online/real time systems require frequent backups within a day. The backup may be incremental or complete. Automating the backup procedures is preferred to obviate operator errors and missed backups. Recovery and business continuity measures, based on criticality of the systems, should be in place and a documented plan with the organization and assignment of responsibilities of the key decision making personnel should exist. An off-site backup is necessary for recovery from major failures/disasters to ensure business continuity. The business continuity plan should be frequently tested.

**D. System and Network Logging:** Operating systems, database packages and even business applications produce a *log* of various tasks performed by them. Most operating systems keep a log of all user actions. Log files are the primary record of suspicious behaviour. Log files alert the administrator to carry out further investigation in case of suspicious activity and help in determining the extent of intrusion. Log files can also provide evidence in case of legal proceedings. The administrator has to select types of information to be logged, the mechanisms for logging, locations for logging, and locations where the log files are stored. The information required to be logged should include login/logout information, location and time of failed attempts, changes in status, status of any resource, changes in system status such as shutdowns, initializations and restart, file accesses, change to file access control lists, mail logs, modem logs, network access logs, web server logs, etc. The log files must be protected and archived regularly and securely.

# 6

# ICT in Education and Health Sectors

---

## 6.1 ICT: The Third Revolution

The adoption and rapid diffusion of information and communication technologies (ICT) has been the most radical technological change of recent decades. Economists have been focusing on the economic consequences of this technical revolution on education, wages, workplace organization, markets and competition, network and innovation for quite some time now.

Telecommunication industry has grown rapidly worldwide. In the developed countries they connect every home and office and in the developing countries they are being rapidly extended through latest wireless technologies to connect homes and offices to the networks and to connect the local network to the global system through optical fiber cables and communication satellites. Information (which when organized constitutes knowledge) is all getting electronified, digitized, coded and stored in vast solid state computer memories. When these are connected to the telecommunication network, information is available to everyone who is connected to the telecom network through appropriate devices like telephone, PCs, TV sets, mobile phones etc. Computers and telecommunication, thus form the information infrastructure for storage, distribution and exchange of information. Internet is now a substitute for broadcasting system, banks, government offices, libraries etc. In short, it is used for all activities involving use of information.

Information and communication technology (ICT) is the base of the third revolution (first being agricultural revolution that lasted for few thousand years and second, industrial revolution that began in the middle of 18th century) in human history. Just as machines of industrial revolution have extended human's physical power, ICTs are extending human brainpower, linking up the brain power of entire humanity in one network. The use of ICTs in every activity of people will make every operation speedier, less expensive and more comprehensive. It will help us to do the things in new ways. It involves development of applications and the deployment of ICT devices for users and the network for connecting them. For example, in Andhra Pradesh, ICT is used for governance and management and delivery of services as well as its diffusion to every enterprise and economic activity of the people to bring about social and economic prosperity. Undoubtedly, those who are educated and skilled and trained to use ICTs would derive more advantages and prosperity than those who cannot (Chowdhary, 2003; Chadha, 2006, 2008).

The growth of ICT has led to the emergence of a 'New Economy' and society is shifting from an era of industrial society into knowledge society. In any knowledge society the growth rate of service sector is higher than that of manufacturing sector. Services themselves are also becoming an increasingly integral part of all production businesses and there are relatively more people working in this sector now (Dahlamn and Utz, 2005).This trend is ultimately leading to an increased emphasis of intangible products. Providing services is a much more challenging task than just delivering goods. Therefore, there will be an increasing focus on making optimal use of resources for maximizing their competitiveness.

## 6.2 ICT and the Knowledge Economy

There has been a long debate in the business information system and economic literature over whether the information technology revolution is paying off in higher productivity and economic growth. The first studies, conducted in 1980s, found no connection between IT investment and productivity at the level of firms, industries or the economy as a whole (Loveman,

1994; Roach, 1987; Strassman, 1990). This is called 'productivity paradox'. This paradox has stimulated economists, management scientists and information system researchers to conduct more rigorous scientific analysis on the relationship between IT and productivity (Brynjolfsosn, 1996; Brynjolfsson and Hitt, 1995, 1996, 1998; Jorgenson and Stiroh, 1999; Oliner and Sichel, 2000; Bosworth and Triplett, 2000; Jorgenson, 2001). These studies used larger datasets and more refined research methods.

The studies revealed positive and significant impacts from IT investment at the firm and country level. Moreover, some of these studies showed that the economic boom and the rise in productivity in late 1990s were largely due to heavy investment in IT and the growth of internet. This may signify that there are substantial adjustment costs in implementing IT and policy makers should not expect dramatic improvements in the growth in the short run. The economic returns come only after a substantial increase in IT investment.

In the studies of impact of IT on the economic growth, IT investment has been broadly defined as the investment in both computers and telecommunications and in related hardware, software and services. Growth of output in firms, industries and the economy may arise from increases in input levels, improvement in the quality of inputs and growth in productivity of inputs. In addition to it, there are complementary factors that influence the payoffs from IT investment. At the firm level, these include organization, management practices (Brynjolfsson, 1996; Bressnahan et al., 2002). At the industry level, they include industry organization and at the national level, they might include economic structure, government policy and investment in human capital (Dewan and Kraemer, 2000).

At the firm level, earlier studies were unable to show the positive role of IT investment in their growth because of inadequate data on IT investment and small sample size (Brynjolfsson and Hitt, 1996; Brynjolfsson et al., 2000). Most discouraging results were found in service firms such as banks and insurance, which showed weak or nonexistent links between IT and productivity (Strassman, 1990; Harris and Katz, 1991).

More recently, Brynjolfsson and Hitt (2000) found that returns to IT investment were not only limited to increase in labour productivity but also in growth of multifactor productivity in the U.S. firms. Gillchrist et al. (2001) in a sample of 1000 Fortune U.S. firms showed that IT had a substantial impact on labour productivity growth in manufacturing companies. They found that in the non-durable goods sector, capital deepening led to increased return to labour productivity. Moreover, these returns are correlated with diffusion and networking of computing throughout the organization. In contrast to it, Lal (2001) did not find a relationship between IT investment and productivity in Indian garment industry. This is consistent with cross country studies by Dewan and Kreamer 2000; Pohjola 2001, wherein they did not find a strong relationship between IT and growth of productivity in developing countries.

Moreover, there is mixed evidence at the firm level as to the impact of IT investment on the financial performance measures such as profitability, partly because the linkage is less direct. While IT investment can directly affect a firm's output; a firm's financial performance is determined by a wider range of strategies and competitive factors that go beyond productivity. Brynjolfsson and Hitt (1996) in a data sample of 370 U.S. firms showed that while IT investment affects productivity and contribute to consumer welfare (through lower prices or better services), it does not necessarily improve profitability.

Several studies show a relationship between IT investment and intermediate measures of operational performance. Barua et al. (1995) in a data sample of 20 U.S. companies found that IT investment affects intermediate measures such as inventory turnover. But he found no evidence that the benefits extended to firms' performance as measured by return on assets. However, Brynjolfsson et al. (2000) found that high investment in IT and organizational capital led to increased market valuation.

While some of the firm-level studies have compared results from different sectors (for example services vs. manufacturing), there is dearth of studies using aggregate data at the industry level. The results are substantially different in different industry sectors. In particular, distinction has to be made between IT-producing and IT-using sectors. The IT-producing sectors are those which manufacture semiconductor, computer, or telecommunication hardware or provide software services that enable these technologies to be used effectively in organizations. The IT-using sectors are the other sectors of economy that apply IT as part of their operations in order to achieve greater efficiency and effectiveness (Dedrick et al., 2003). They include sectors as manufacturing, wholesale and retail trade, finance, insurance and real estate, business and professional services and so on.

There is a considerable agreement among economists that productivity has increased in IT-producing industries (Gordon 2000; Oliner and Sichel, 2002). Jorgenson (2001) attributes two-third of productivity growth during 1995-99 to the IT-producing sector. Specifically, he attributed the growth in productivity to continual technical innovation (R&D) in semiconductor and computer industries owing to which the computer and telecommunication equipment prices declined by 90 percent during the 1994-99. The increase in productivity in the IT-producing sector has naturally contributed to the multifactor productivity growth in the whole economy given its increasing share of the national economy.

On the other hand, there is some debate about whether productivity growth has also occurred in the IT-using industries. Most studies have attributed growth in productivity in late 1990s to the IT-using industries as well as IT-producing industries (Whelen, 2000; Baily and Lawrence, 2001; Council of Economic Advisor, 2001). However, Gorden (2000) found no evidence of productivity acceleration outside IT-producing and durable goods manufacturing industries.

Bosworth and Triplett (2000) in the study of service sector (IT-using industries) found that increase in labour productivity was due to growth in multifactor product, IT capital deepening and increased use of outsourcing. However, they concluded that multifactor productivity was dominant factor contributing to the productivity acceleration in service industries during late 1990s. Van Ark et al. (2004) concluded that better economic performance in the USA relative to Europe can be partly attributed to faster growth in IT-using service sectors such as wholesale and retail trade and financial brokerage. He stressed the importance of accumulation of intangible capital, knowledge and skilled labour in IT in explaining growth in productivity in services in USA.

In developing countries, however, it is difficult to derive systematic evidence of correlation between ICT use and economic performance. Some case studies are available in traditional export oriented industries in these countries. While the traditional export sectors are engaged in extensive use of e-mail, they are not yet really connected to emerging e-market places. At the same time, however, there is evidence of improved market access and rapidly increasing exports by selected developing and transition economies. This implies that there are promising ICT related growth prospects in global markets for developing countries. Udo and Edoho (2000) in his study on ICT initiatives in Africa describes various institutions that have attempted to foster diffusion of ICT throughout the continent.

Many of these initiatives are public-private partnerships and some are non-profit organizations. Each is actively involved in deployment and integration of ICT. He finds that India has been successful in extending the information systems to rural areas. Lal (2002) examined comprehensive data from 51 Indian firms on numerous aspects of performance including data on ICT investment, wages, exports, profits etc. He concluded that firms that adopted more advanced e-business tools generated higher levels of exports. Thus, it appears that the adoption of sophisticated e-business technologies may improve economic performance.

Dutta et al. (2003) maintained that China has become a highly competitive producer of wide range of ICT products, from basic electronics and PCs to mobile telephones and high-end semiconductors. The author asserts that by adapting itself to membership in the WTO, China has expanded export-oriented ICT production facilities, as well as to local consumption. The rapidly improving skills of local producers, users and development of science parks have contributed to China's enhanced productivity and upgrading of ICT production. China has also contributed to the growth of the world economy through its increased demand for imported ICT products and services.

Moodley (2002) conducted an in depth quantitative and qualitative analysis of the use of business to business e-commerce on 120 manufacturing firms in South Africa. His evidence indicates that the incidence of IT use is fairly low. Although 87 percent firms had access to internet, only 49 percent of firms had comparable website and only 22 percent were using internet for order taking. He states that business to business e-commerce is in the early stages of its evolution. A study by McKinsey (2004) on the prospects for IT services and software sector in Armenia reveals that this sector started to develop since the second half of 1990s and accounted for 2 percent of the economy. At the same time, the productivity here is at the level of half of that in comparable sectors in India. The main impediments to continued progress are the connectivity, lack of supporting environment, monopolistic position of foreign owned telecom firms and a right set of government policies.

Thus, the empirical findings on IT and economic performance have found a positive correlation between some proxy for IT investment and some proxy for economic performance at firm, industry and country level. Furthermore, complementary investment in IT-related labour and organizational factors (improvement in communication, networking and better coordination made possible by ICT) provide environment for maximizing the returns on IT investment that lead to improvement in productivity growth. The evidence suggests that IT investment will have sustained, long-lasting impact on productivity and economic growth if government policies facilitates faster diffusion of information technology and better allocation of resources.

### 6.3 ICT for Poverty Alleviation

The ultimate objective of planned development is to ensure human well-being through sustained improvement in the quality of life of the people, particularly the poor and the vulnerable sections of the population. In terms of policy measures, it requires emphasis on what in economic parlance is called "Social Sector Development". Policies that aim at human development ultimately contribute to sustained growth as a healthy, educated and skilled work force can participate more effectively in the process of economic development.

Over the past few years, several initiatives have demonstrated the potential to use ICT for achievement of development goals such as poverty alleviation, increased access to education, health services and reduce gender inequalities.

Alleviation of poverty has been a priority issue for the policy makers in India. Large amounts have been allocated in the various budgets year after year for eradicating poverty. ICT can impact poverty through five channels as discussed in the literature. These channels are:

1. The ICT enables the poor to have access to information on government policies.
2. The ICT increases the connectivity to the market, which will lead to realization of economic benefits in terms of getting suitable prices for rural produce and also creating

employment opportunities.

3. The ICT facilitates better monitoring of public administration, social services and development programs.
4. The ICT increases productivity and extends the sphere of economic activity in rural areas. For example fisherman can access information on the location of fish shoals for more successful fishing expeditions.
5. ICT contributes to the women empowerment and helps in the areas of healthcare, childcare, nutrition and local indigenous medicine.

There have been a number of isolated attempts in different states by both private and government organizations to use ICT for betterment of rural population in India. The Gyandoot project, implemented in the Dhar district of Madhya Pradesh, the Community Information Centres run by the Department of Information and Technology in North-Eastern States, and e-Chaupals run by Indian Tobacco Company (ITC) in different states can be stated as functional example of using ICT for rural development in India (Government of India, 2002b, 2008; NASSCOM, 2005). From an examination of the examples quoted here, it is possible to suggest a framework of ICT implementation that is engaged with relevant processes and principles of poverty alleviation. The framework so derived facilitates an understanding of how ICT can help alleviate poverty. A pro-poor ICT policy begins with a development commitment that targets poverty alleviation, with government acknowledging its role as a major employer and user of ICTs. ICT policies should be concerned both with ICT production and with ICT use.

The concern here is for policies for ICT use that specifically target poverty alleviation, as opposed to e-commerce, e-government, e-learning, and the like. So far, the use of ICTs for poverty alleviation has not achieved the status of a mass movement, despite the many promising results that have emerged from experimental installations.

## 6.4 ICT for Education in India

In the past 15 years, there have been profound changes in the country and around the world. India is transforming itself into a knowledge-based economy—an economy that creates, disseminates and uses knowledge for its growth and development. The knowledge economy is often taken to mean only high-technology industries or information and communication technologies (ICTs). However, in broader terms, it means how any economy uses new and existing knowledge to improve the productivity of agriculture, industry and services. In the current knowledge driven era, India has been a leading knowledge society. A knowledge society is broader than a knowledge economy and it includes spiritual, social, intellectual and philosophical knowledge.

A knowledge-based society helps in finding new and innovative ways to meet the challenges of building a just and equitable social order and seek urgent solutions. High-tech science and technology, research and technology, linkages between technical development and business, growth of IT sectors, education and health sectors, biotechnology sectors and manufacturing sectors and growth in services are critical areas in knowledge economy of India.

The success of IT industry overall influenced competitiveness in other sectors as well, by building confidence in Indian industry. India is undertaking new work in financial analysis, industrial engineering, analytics, and drug research among others. The increasing use of information technology in Indian economy has made it one of the fastest growing economies in the world. In fact, Indian economy is the fastest growing economy in terms of spending on IT in the entire Asia-Pacific region (Government of India, 2008).

Technology may help towards building equitable knowledge based society by facilitating better access to education in remote locations or people from underprivileged sections of society in an economical manner. National level education policies have achieved some success in raising the level of education and literacy in India.

The fast growing IT sector has created huge career opportunities and has a profound impact on socio-economic dynamics of the country. Further, it is seen that industry is changing from BPO to KPO (knowledge process outsourcing) with increased need for knowledge professionals. To retain our position as a major player in IT arena various projects in the area of curriculum development, content development and course material development, faculty training, e-learning, online examination system etc. have been implemented.

To become a knowledge economy India need to keep developing its human capital base and creating knowledge worker. Emphasis has been laid on expansion with equity, use of information and communication technology and promotion of quality education.

The National Mission on Education through ICT aims to extend computer infrastructure and connectivity to over 18000 colleges in the country including each of the departments of nearly 400 universities/deemed universities and institutions of national importance as a part of its motto to provide connectivity up to last mile.

Knowledge has been recognized as the key driving force in the 21st century and India's ability to emerge as a globally competitive player will substantially depend on its knowledge resources. Well-educated and skilled people are essential for creating, sharing, disseminating and using knowledge effectively. The knowledge economy demands a new set of competencies, which include not only ICT skills but also soft skills such as problem solving, analytical skills, group learning, working in a team based environment and effective communication.

Fostering such skills requires an education system that is flexible and develops core skills that encourage creative and critical thinking. China's experience in this area is instructive. Its emphasis on secondary education has provided a firm basis for global expansion of manufacturing activities. Investment in basic education is fundamental for a country to improve its productivity and quality of its labour needed for development. Investment in IT by education sector is likely to be focused mainly on basic school education in India.

Education is one area, which has been universally accepted to have the potential to reduce the digital divide. Information and communication technologies are ushering unprecedented changes in nature content and imparting of education. They have opened new vistas for teachers, students, parents and public in education by making them more involved. The governments across the world have recognized the importance of ICT and have taken appropriate policy measures for ICT enabled education. Developed and developing countries have begun to spend considerable resources on this suite of technologies to improve the quality of education.

In its size and diversity, India has the third largest higher education system in the world, next only to China and the United States. Education in India has received enormous financial support from both the Central and the State Governments.

### 6.5 ICT for Health Services

Information technology is poised to revolutionize healthcare through new ways in human connectivity. The healthcare sector comprises of, health insurance companies, hospitals, physician organizations, nursing and personal care facilities, pharmaceuticals, employees and government health organization. The healthcare industry is attempting to manage and reduce rising medical costs. Investment in IT by healthcare sector is likely to be focused mainly on customer (patient) management systems and maintaining electronic medical records. For instance, the US regulatory provisions (Health Insurance and Portability and Accountability Act) requires healthcare service providers to maintain an increasing portion of their medical records in electronic form, to enable patients to switch service providers easily.

Health of a nation is an essential component of development, vital to its economic growth and internal stability. Assuring a minimal level of healthcare to the population is a critical constituent of the development process. Considerable achievements have been made since Independence in 1947 to improve health standards such as increase in life expectancy, decrease in infant and maternal mortality, and eradication of small pox and guinea worm. Nevertheless, problems abound. Malnutrition affects a large proportion of the population especially women and children. An unacceptably high proportion of the population continues to suffer and die from new diseases apart from the existing ones. Pregnancy and childbirth related complications also contribute to the suffering and mortality of women.

There is a strong link between poverty and ill-health. Ill health creates immense stress even among those who are financially secure. Onset of a long and expensive illness can drive the economically well-off into poverty. The importance of public provisioning of quality healthcare to enable access to affordable and reliable heath services cannot be underestimated. This is specially so in the context of preventing the non-poor from entering into poverty and reducing the suffering of those who are already below poverty line.

Public institutions played a dominant role in the Indian healthcare sector in the past, in urban as well as in rural areas. However, the public healthcare system has been on a decline during the last two or three decades because of non-availability of suitable medical and para-medical staff, diagnostic services and medicines. Equally distressing is the fact that essential therapeutic drugs are not supplied in most public health institutions with the exception of some States. Consequently, there has been a decline in the percentage of cases of hospitalized treatment in government hospitals and a corresponding increase in the percentage treated in private hospitals. Despite higher costs in the private sector, this shift shows people's growing lack of trust in the public system. Critical shortage of health personnel, inadequate incentives, poor working conditions, absenteeism, long wait, inconvenient clinic hours, insensitivity to local needs, inadequate planning, management, and monitoring of services/facilities appear to be the main reasons for this shift.

India has to deal with rising costs of healthcare and growing expectations of the people. The challenge of quality health services in remote rural regions has to be urgently met. Given the magnitude of the problem, there is the need to transform public healthcare into an accountable, accessible and affordable system of quality services. Special attention needs to be given to the health of vulnerable groups like adolescent girls, women, children, older persons, disabled and primitive tribal groups. The contribution of the private sector in providing primary, secondary and tertiary services should be enhanced through various measures including partnership with the Government. Good governance, transparency and accountability in the delivery of health services should be ensured through involvement of PRIs, community and civil society groups.

Rising income levels and the emergence of health insurance has facilitated the growth of high quality healthcare sector in India although the capacity to pay for health services remains limited to middle and high income brackets. With the opening of the insurance market in 2000 to private players, health and medical insurance has shown significant growth. Cashless claim facilities offered by many players have led to a significant rise in health insurers.

The public healthcare system is not able to meet the growing demands for healthcare facilities. The gap between demand and supply has to be bridged by the private players, and this has facilitated the growth of the corporate healthcare industry in India. Emergence of strong corporate players in healthcare is not a substitute for expansion of public sector health facilities but it has led to professionalization of the industry, better hospital management and higher investments in infrastructure facilities.

**6.5.1 Towards e-Health:** Eleventh Five Year Plan (2007-12) had ambitious programmes to use information technology for health services and usher India into the age of e-health. To quote, "Appropriate use of IT for an enhanced role in healthcare and governance will be aimed at during the Eleventh Five Year Plan. It is feasible to set up a National Grid to be shared by healthcare providers, trainers, beneficiaries and civil society. The country already has the advantage of a strong fibber backbone and indigenous satellite communication technology with trained human resources in this regard. A number of pilot projects on e-Health over the past years by private concerns, corporate, NGOs, medical colleges and research institutions have been set up. The successful outcome of many of these initiatives needs to be evaluated and scaled up.

Health Management Information System (HMIS) would be an important new initiative utilizing developments in the field of information technology. A computerized web enabled data capturing and analytical system will be established to provide valid and reliable data and reports for use at all levels. This would not only facilitate proper monitoring and evaluation of different programmes under implementation but will also help in various aspects of service delivery. The HMIS will also integrate the various vertical systems having their own reporting machinery into an integrated umbrella of holistic monitoring and evaluation to cater to the needs of Sarva Swasthya Abhiyan. The data will flow directly from the periphery. The Integrated Disease Surveillance Project (IDSP) will eventually be a by-product of the HMIS. As the system stabilizes and the penetration of computerization at the Block level increases, the system will be modular enough to expand the scope to the remotest areas. Wastage of drugs due to date expiry also needs to be curtailed by demand driven management and redistribution of medicines nearing date of expiry. Health Management Information System (HMIS) when fully developed and implemented will track demand and supply and continuously monitor the drug situation.

Telemedicine could help to bring specialized healthcare to the remotest corners of the country. Telemedicine is likely to provide the advantages of tele-diagnosis, especially in the areas of cardiology, pathology, dermatology and radiology besides continuing medical education. It will be of immense use for diagnostic and consultative purposes for patients getting treatment from the secondary level healthcare facilities. The efficacy of telemedicine has already been shown through the network established by the Indian Space Research Organization (ISRO), which has connected 42 super-specialty hospitals with 8 mobile telemedicine vans and 200 rural and remote hospitals across the country through its geostationary satellites. So far about 3 lakh people have benefited from this programme. Facility of telemedicine will be provided in district hospitals and Government Medical Colleges.

The e-Health initiatives to be taken up during the Eleventh Five Year Plan are:

- Training, Education and Capacity Building for e-Health.
- Monitoring by e enabled Health MIS (HMIS) to ensure timely flow of data and collation to be used at various levels.
- GIS Resource Mapping of various health facilities (Allopathic and AYUSH), Laboratories, Training Centres, Health Manpower and other inputs to optimize utilisation
- Providing service delivery and other e-enabled activities like, disease surveillance, tele-consultations, health helpline, district hospital referral net and e enabled mobile medical units." (Government of India, Planning Commission, *Eleventh Five Year Plan* (2007-12), Vol., II, pp. 85-87.)

### 6.6 ICT and the Manufacturing Sector

The potential growth areas in Indian IT using industry

includes manufacturing, telecommunications, banking and financial services, utilities, retail and wholesale trade, and healthcare. These sectors have been traditionally low spenders on IT but are now increasing IT spends due to regulatory and market pressures (NASSCOM, 2003, 2008). The level of IT investment varies across the sectors depending upon level of competition and expectations regarding profits and growth. To look into the growth effects of ICT industry, we have selected one knowledge intensive sector- drugs and pharmaceuticals and some critical sectors of knowledge economy of India like education, health and manufacturing.

Manufacturing sector comprises of discrete and process manufacturing. In discrete manufacturing, products are produced through the assembly of distinct and/or individual parts. It includes sub-industries like apparel and textile products, leather and leather products, fabricated metal products, automobile industry, auto-component sector and fast moving consumer goods industry. In process manufacturing, products are created primarily through one or more continuous activities including milling, curing, weaving, smelting and refining. It includes following sub-industries: textile mill products, primary metal industries, heavy and light engineering and miscellaneous process manufacturing (Gupta et al., 2007).

While the earlier technologies favoured mass production of a given model of a product, the current micro-electronics and information technologies support flexibility in designs, model and product mix. The modern manufacturing methods characterized by computer aided designs, computer aided manufacturing and flexible manufacturing systems encourage repeated changes in product specifications. Studies on Indian small and medium enterprises manufacturing garments, electrical and electronic goods (Lal, 1996, 1998) show that introduction of information technology results in a shift in manufacturing configuration. Evidence indicates that IT promotes productivity, profits and exports. The advent of higher-profits in-turn makes it possible for the firm to invest more on information technology.

In Indian manufacturing sector, IT spending increased from ₹ 3,382 crore in 2001 to ₹ 12,498 crore in 2008 and accounted for 14 percent of total domestic IT market spending in 2008 (Gupta et al., 2007). To analyze the effect of IT spending on the growth of manufacturing sector, we have regressed profitability on export intensity, advertising intensity and ratio of IT spending to total cost in manufacturing sector for the period 2001-08.

# 7

# ICT in Agriculture and Rural Development

---

## 7.1 Need for IT-based Services to Farmers

Indian farmers, for whose welfare a huge government machinery is devoted, still suffer from the absence of right information at the required time. The farmers require timely information on weather conditions, sowing time, availability of inputs including credit, expert advice on maintaining crops in healthy conditions, information on markets and on all other areas of interest to them.

The rapid strides that the country has registered in the IT field will remain incomplete unless IT is fully utilised to ensure more efficient and productive Indian agriculture. Practically, there is no area in agriculture in which IT has no role to play. The immense scale and diversity of Indian agriculture provides the ultimate challenge to the potential of information technology.

The adoption of IT services in agriculture depends on the main functionaries involved in Indian agriculture. These functionaries can make a big contribution to the growth of agriculture with the assistance of IT. For this purpose, the following functionaries can be considered: (a) farmers, (b) industries providing inputs to agriculture, (c) industries dealing with agricultural output, (d) Central and State Governments, and (e) NGOs working for the benefit of farmers and agricultural universities and research centres.

In spite of the best efforts and expenditure, the conventional apparatus has not been able to deliver the goods satisfactorily. Herein lies the role of IT which can efficiently address the concerns of farmers stationed at even remote locations. Low literacy levels, cost of computers, poor communication infrastructure make it impossible for individual farmers, particularly small farmers, to directly adopt IT. This calls for institutional efforts to provide IT-based services to farmers.

In order to keep pace with the state-of-the-art technologies, National Informatics Centre (NIC) has been conducting various training programmes on IT application on a regular basis from time to time for the user organizations of agriculture sector. Agricultural Informatics Division of NIC has taken up various initiatives in bringing IT-led development which includes web-enabled applications, GIS-based applications, multimedia applications, database applications and E-governance and training.

India has emerged over the last decade as the most preferred destination for outsourcing of ICT services. The vibrant ICT industry is contributing immensely by providing information about latest technology and international business practices. Various sectors of the Indian economy such as industry, finance, insurance, communications and transport have adopted information technology in a big way. However, agricultural sector of the economy is lagging behind in utilizing ICT services. Some efforts have been made in this regard but they are by no means adequate. The rapid strides that the country has registered in the IT field will remain incomplete unless IT is fully utilised to ensure more efficient and productive Indian agriculture. Practically there is no area in agriculture in which IT has no role to play. The immense scale and diversity of Indian agriculture provides the ultimate challenge to the potential of information technology.

## 7.2 IFFCO-ISRO Cooperation

IFFCO has initiated IT-based services for farmers and cooperative societies. It has taken up a project in association with Indian Space Research Organisation (ISRO) to utilise satellite based remote sensing data and Geographical Information Systems (GIS). Developed countries have been utilising precision farming with the help of IT tools for a long time. While this will take a long time for our country due to small holdings, it is to be noted that GIS has an invaluable role to play even in the existing conditions. Remote sensing and GIS information can provide warnings on evolving crop stresses, crop vigour etc.

The IFFCO-ISRO GIS project extends support for efficient and timely availability of IFFCO's fertiliser to farmers though better logistics and efficient operations. Based on the experience gained from this project, more intensive services based on GIS are envisaged for the entire country. In addition to the GIS-based services, effort is being made to create databases that contain information of interest to the farmers. These include recommendation on package of practices for major cereals, pulses, horticulture, floriculture and animal husbandry etc. Information on various inputs such as seeds, fertilisers, credit is provided. An important service envisaged is to provide access to the nearest expert in case of stress or any other problem witnessed in the crops. Facilities are sought to be provided to encourage and share farm experiences by forging various crop forums. Many of the agricultural extension services are also proposed to be made online using multimedia facilities.

In order to encourage farmers to obtain best possible price, information regarding prices of various crops prevailing in approachable markets (mandies) is also being provided. Other areas of interest to farmers such as distance education, location specific news etc. are also planned.

### 7.3 Cyber Dhabas

To take IT to the door steps of the farmers, it is proposed that a large number of access points called cyber dhabas are setup. These cyber dhabas should be provided with a PC and a modem with telephone connection to connect to the databases. The farmers can take the assistance of the operator of cyber dhabas for a nominal charge. The major bottleneck in spreading e-culture to rural areas is related to connectivity i.e. ensuring that the access points can get connected to the databases which are in selected locations. Since dialup lines are very slow, other viable options are required to be explored. Unfortunately, the alternatives are expensive and may not be feasible. One silver lining is that several private operators are connecting important cities with fibre optic which provides a very reliable and fast access. Since these will pass through rural areas, it is possible to explore the possibility of tapping this potential by laying the last mile connectivity. Once this is done, substantial segment of rural India can access the IT-based services.

### 7.4 E-Choupal

E-Choupal offers farmers the information, products and services they need to enhance productivity, improve farm-gate price realisation, and cut transaction costs. Farmers can access the latest local and global information on weather, scientific farming practices, as well as market prices at the village itself through this web portal-all in Hindi. E-Choupal also facilitates the supply of high quality farm inputs as well as the purchase of commodities at the farm.

Given the literacy and infrastructure constraints at the village level, this model is designed to provide physical service support through a choupal sanchalak, himself a lead farmer, who acts as the interface between the system and the farmers. The contents of this site in their entirety are made available only to the registered sanchalaks.

### 7.5 Kisan Call Centres (KCCs)

The scheme was launched on 21stJanuary 2004 to provide agricultural information to the farming community through toll free telephone lines. A country wide common eleven digit number '1800-800-1551' has been allocated for KCC. The replies to the queries of the farming community are being given in 22 local languages. Calls are attended from 6.00 am to

10.00 pm on all 7 days of the week. In order to make farmers aware of this facility, audio and video spots on Kisan Call Centres are being broadcast telecast through All India Radio/ Doordarshan and private television channels. Kisan Knowledge Management system (KKMS) to provide correct, consistent and quick replies to the queries of farmers is being developed for each State/Union Territory.

### 7.6 National e-Governance Plan in Agriculture (NeGP-A)

The underlying vision behind this project is to "create an environment conducive for raising farm productivity and income to global levels through provision of relevant information and services to the stakeholders." The initial phase of the project was envisaged to identity the key elements that need to be focused in agriculture and allied sectors. The next phase focused on detailed strategy, road map and guidelines for implementation of E-governance in agriculture at both Central and State levels. There are various IT initiatives/schemes undertaken or implemented by DAC which are aimed at providing information to the farmers on various activities in the agriculture value chain. These initiatives need to be integrated so that farmers may be able to make proper and timely use of the available information. Such information is intended to be provided to farmers through various channels including Common Service Centres, Internet Kiosks and through SMSs.

### 7.7 Other IT Services to Agricultural Sector

**7.7.1 LAN at Ministry of Agriculture:** A high speed Local Area Network was established in the Ministry of Agriculture spread over various buildings viz. Krishi Bhawan (~500 nodes), Shastri Bhawan (~200 nodes), Krishi Anusandhan Bhawan (~100 nodes) etc. In addition, Agricultural Informatics Division has extended its full support (round the clock operation) to the Ministry of Agriculture at the time of super Cyclone in Orissa and Gujarat Earthquake.

Video Conferencing facilities were used extensively from the Krishi Bhawan studio to Orissa and Gujarat on continuous basis.

**7.7.2 Plant Protection Informatics and Communication Network (PPIN):** The PPIN envisages to link the Directorate of Plant Protection Quarantine and Storage (DPPQS), Faridabad, National Plant Protection Training Institute, Hyderabad, CIMPS, LCC, Central Pesticides Testing Laboratories, and State Pesticides Testing Laboratories and to disseminate information related to plant protection and pesticides. As a part of project, the computerization of the CIB/RC of DPPQS has been taken up by NIC.

**7.7.3 Agricultural Marketing Information Network (AGMARKNET):** NIC has taken up an ambitious project "Agricultural Marketing Information Network (AGMARKNET)" on turnkey basis for the Directorate of Marketing and Inspection (DMI), Department of Agriculture and Cooperation, Ministry of Agriculture. The project aims at improving the prevailing agricultural marketing information system by minimising the gap between generation and dissemination of market information. The major components of AGMARKNET are the following: (a) establishment of computing facilities and networking, (b) development of human resource, (c) information transmission, (d) development of database and (e) portal on market information. NIC has tied up with Bharat Sanchar Nigam Ltd. (BSNL) to provide internet facilities at the AGMARKNET nodes.

**7.7.4 Strengthening of Informatics in the Offices and Field Units of the Department of Agriculture and Cooperation (DAC)–DACNET:** To promote e-Governance in Agriculture at the Centre and to provide support to States/UTs for the same, DAC is implementing this Central Sector Scheme. The scheme has the following components:

1. IT apparatus at DAC Headquarters, field offices and Directorates.
2. Development of Agricultural Informatics and

Communication.

3. Strengthening of IT Apparatus in Agriculture and Cooperation in the States and Union Territories (AGRISNET).
4. Kisan Call Centres.

**7.7.5 MIS Support to Food Processing Industries:** To cater to the needs of the Department, NIC has developed various information systems viz. FPO and licensing monitoring, Plan scheme monitoring, industrial approvals, processed food products, FPO on-line status, industrial entrepreneur's memoranda query information, country profile and international cooperation etc.

**7.7.6 Agricultural Extension Information System Network: VISTARNET:** Research, education, extension, and training are considered as four pillars of sustainable agriculture. Generation and transfer of technology have become very crucial the world over. As a step towards making technology reach the small holders (resource-poor-farmers), efforts are being made to establish VISTARNET–NICNET based Agricultural Extension Information System Network, in India, linking extension functionaries at Central, State, and District level. The required funds for implementing VISTARNET will be provided through the National Agricultural Technology Project (NATP). As a part of VISTARNET, Informatics Development for the Directorate of Extension has been taken up by NIC.

**7.7.7 Indian Agriculture On-line:** NIC has submitted an IT-Plan for the agriculture sector with the objective of establishing NICNET based Agricultural Informatics and Communication Network (AGRISNET) to facilitate higher and sustainable agriculture productivity and also putting Indian Agriculture on-line in the Country.

**7.7.8 Infrastructural Facilities and Services Provided by NIC to the Ministry of Agriculture:** These are the following:

1. INTRANET/INTERNET.

2. Telecommuting programme.
3. Video conferencing.
4. Information kiosk.
5. In-house training facilities.
6. Web site design and development.
7. Web enabled applications.
8. Information bulletin.
9. Intranet applications.
10. Agriculture portal.
11. NICNET based Public Information and Facilitation Centre.

**7.7.9 Animal Production and Health Information Network (APHNET):** APHNET envisages building up of reliable databases and network based information systems for all activities of the animal husbandry and dairying sector at district, state and national level, using NICNET facilities. In view of importance of making technology reach the small holders through IT, the plan also suggests to establish APHNET nodes at about 42,000 veterinary polyclinics, hospitals, and dispensaries and veterinary aid centres for strengthening animal disease surveillance and advisory system in the country. As a part of APHNET, National Project on Rinderpest Eradication (NPRE) which aims at capturing animal disease related information from various State Animal Husbandry Departments has been entrusted to NIC.

**7.7.10 Agricultural Research Information System Network (ARISNET):** The ARISNET, a close user group of institutions in National Agricultural Research System of India (NARS) on NICNET, links ICAR Institutions and their Regional Research Stations, Central Agricultural University, State Agricultural Universities and their colleges, Krishi Vigyan Kendras, Zonal Research Centres and Project implementing centres with the ICAR Headquarters. ARISNET has become an integral part of agricultural research, extension and education process.

**7.7.11 Market Information System for Horticulture:** This project has been taken up by NIC on a turnkey basis. As part of the project, NICNET based internet/intranet facilities have been established at National Horticulture Board (NHB), Gurgaon. Computing facilities have been created at 33 market centres of the Board located all over the country. NIC has developed and implemented the necessary software for evolving a comprehensive database of the prices and arrivals of fruits and vegetables being received by NHB headquarters from 33 market centres on a daily basis. This information is being used by the Ministry of Agriculture, its related agencies and markets through National Bulletin on fruits and vegetables.

**7.7.12 Integrated Fertilizer Management Information System (FERMIS):** The Department of Fertilizers (DOF) in collaboration with National Informatics Centre (NIC) has introduced computer-based methods for decision support with the major objective to evolve an evaluation system which ensures a uniform system of Planning and control mechanism with signalling system to highlight deviations from desired performance indicators by Plants/organisations for all the public sector fertilizer enterprises. Thus, an Integrated Fertilizer Management Information System (IFMIS) of worth ₹ 1.99 crore has been executed through NIC and various IT based systems developed covering Planning, Movement, Import, Handling-Payment, Project Monitoring, Performance Monitoring and Evaluation aspects to strengthen Fertilizer Informatics for decision support. Information exchange for decision support has been promoted through Fertilizer Informatics Network (FERTNET) extended to Department of Fertilizers and various Fertilizer Companies viz. IFFCO, NFL, MFL, GNFC, FCI and HFCL.

**7.7.13 Activities in the North-Eastern Region:** Under the IT apparatus at the Field Offices and Directorates of DAC (DACNET), the following offices in the North-Eastern region have been covered:

1. Directorate of Marketing and Inspection at Guwahati and Shillong.

2. Central Integrated Pest Management Centres at Guwahati, Aizawl, Dimapur and Gangtok.
3. Regional Bio-fertilizer Development Centre at Imphal.
4. North-Eastern Regional Farm Machinery, Testing and Training Institute, Sonitpur, Assam.
5. Support under AGRISNET.

### 7.8 Computerisation of Agricultural Census

Recognizing the predominance of the agriculture sector in the Indian economy, Department of Agriculture and Cooperation collects and maintains agricultural statistics such as number, area, tenancy, land utilisation, cropping pattern and irrigation particulars of different classes of operational holdings regularly and make it timely accessible to the planners and policy makers for decision making. This project was entrusted to NIC and it made significant headway in almost all the States/UTs except Bihar, where Census is yet to be conducted. The salient features are the following:

1. Creation of large database of about 8000 million bytes at National Level and 1 GB at state level.
2. Processing and tabulation of Agricultural Census and Input Survey database at District/State/National level.
3. Development and implementation of information retrieval system at micro and macro level for decision making at various levels.
4. Selective information will be put on INTERNET/INTRANET for easy access of information to public to ensure transparency in the Government functioning.
5. Training programmes on software implementation for officers of States/UTs and Central Government.

### 7.9 Rural Portal of Ministry of Rural Development (MoRD)

MoRD has developed a number of applications to automate the processes of information generation and strengthening e-governance. Efforts have also been made to develop national level ICT solutions for some important activities like the land records computerization, MIS for Mahatma Gandhi National Rural Employment Guarantee Act (MGNREGA) and, online monitoring of schemes like Rural Bazaar, RuralSoft.

Rural Portal provides a gateway to around 150 websites. It provides details of schemes of the Ministry, their physical and financial status, release position, sanction orders etc. Apart from this, news related to Ministry is also uploaded daily onto the site. Circulars of public interest, information regarding events, tenders/notices, e-mail are also uploaded regularly onto the site. Online SARAS Mela is available on the website to web showcase the products made by rural artisans for wide publicity and to create national/international market for their products. Applications have been developed to monitor on-line the Monthly Progress Reports of schemes of the Ministry.

NIC-DRD informatics cell has designed and developed the software for the household survey BPL Census 2002. The software was designed in such a way that it uniquely identifies a family and all the members of the family at national level. It also assigns a unique number to all panchayats which are the executing agencies/service providers for most of the schemes of MoRD. The family and the persons whose information is captured in the software are the beneficiary/service receiver for most of the schemes of MoRD and are uniquely identified.

## 7.10 National Rural Employment Guarantee Act Software Package (NREGASoft)

MoRD and NIC commenced the preparation of appropriate E-governance solution to strengthen NREGA in October 2005 and when the scheme was launched in February 2006, the NREGASoft was also launched across the country. The software is available to all stakeholders online through http://NREGA.nic.in and also could be downloaded for off-line working. The package is Unicode-enabled and supports local languages. The training to use the software has been organized in States.

NREGAsoft is web-based software prepared to capture all the activities under NREGA at National/State/District/Block and Panchayat level. Using this portal each stakeholder can input information and access his/her own information through this portal. NREGAsoft captures Registration of workers, work demanded, number of days of employment for whole family, funds transferred/utilized and number of works undertaken under the scheme. Various stake holders of the project are the following:

1. Citizens.
2. Gram panchayats, block panchayats zilla panchayats.
3. Workers.
4. Programme officers.
5. District programme co-ordinators.
6. Implementing agencies other then PRIs.
7. State RD Departments.
8. Ministry of Rural Development and administrators in Government of India.

**7.10.1 Various Modules of the Software:** These are the following:

**A. Beneficiary Management Module:** It captures registration, demand for work, work allocation and muster rolls on which a person worked, the software has the provision of payment of wages through bank/post office as it captures the bank/post office account number, bank/post office name, branch name for all the person who demanded job and their account number are shown in muster roll against their name and calculate unemployment allowance, if any and also keep tracks of number of days of employment of a family.

**B. Fund Management Module:** It captures the funds transferred from MoRD/States to districts and then to programme officers/panchayats and expenditure incurred by various implementing agencies on labour, material and contingency. Hence, it keeps track of each and every paisa spent under the scheme.

**C. Works Management Module:** It captures information about the various works under taken under the scheme at various level. It facilitates online approval of projects and keeps track of time taken for approval of project. Each project is provided a unique ID and status of work is maintained in the system.

**D. Grievance Redressal System:** It allows a worker/citizen to lodge complaint and trace the subsequent response.

**E. Staffing Position Module:** It captures name, telephone numbers etc. of all the officials, planning and implementing agencies from gram panchayat to MoRD involved in NREGA, thus strengthening communication and coordination among them.

**F. Alerts:** The software also gives alerts to implementing agencies about the various irregularities, important activities, messages for funds to be received by the agencies.

**G. Gram Panchayat Accounting System:** NREGA put lot of responsibility on gram panchayats as registration, demand for work and allocation of work can be done only by gram panchayats. Software assists gram panchayats as follows:

1. It keeps track of 100 days of the employment to a family.
2. It ensures that a worker should not work on two worksites simultaneously.
3. The software keeps track of the fund released from the MoRD/States to the districts and then from districts to blocks/gram panchayats so it helps in reconciliation of accounts.
4. It provides status of available funds in the accounts at panchayats, blocks and districts levels.

To sum up, proper planning, monitoring and corrective action requires availability of timely information of right quality. The Central and State Governments require to hasten the process of computerisation of all the records and transactions. Since the government has a major say in promoting agriculture, this will have cascading effect on the entire rural areas. To begin with, it is necessary that the government agencies should accept conventional reports in the electronic format. This will bring about much desired speed and effectiveness in governance. Many State Governments are showing interest in developing proper systems for e-governance. But the implementation of the same is still requires a lot to be desired. The Central and State Governments should also make information available for ready reference by those interested. This will be not only helpful to people but also will raise interest levels in IT-based services. Similarly, research institutions and NGOs should exchange and provide information through a well conceived network. These measures will encourage end user, i.e. farmer's participation and will provide the much needed input from them.

The face of the Indian agriculture can be transformed by a well conceived deployment of IT. The potential of IT as yet remains untapped and urgent measures are required to derive maximum benefit. The key players involved in this process such as industry, government and educational institutions and research centres are required to make contributions in this endeavour. The initiative to develop necessary IT-based agricultural services need to be developed immediately. Parallel steps to develop necessary IT communication infrastructure are to be taken up along with the utilisation of fibre optic network wherever it is passing through the rural segments.

The Karnataka Government's Bhoomi project has led to the computerisation of the centuries-old system of handwritten rural land records. Through it, the revenue department has done away with the corruption-ridden system that involved bribing village accountants to procure land records; records of right, tenancy and cultivation certificates. The project is expected to benefit seventy lakh villagers in 30,000 villages.

A farmer can walk into the nearest taluk office and ask for

a computer printout of his land record certificate for ₹ 15. He can also check details of land records on a touch-screen kiosk by inserting a two-rupee coin. These kiosks, installed at the taluk office, will provide the public with a convenient interface to the land records centre.

# 8

# ICT for Financial Services

Information technology (IT) has transformed the functioning of businesses the world over. It has bridged the gaps in terms of the reach and the coverage of systems and enabled better decision-making based on latest and accurate information, reduced costs and overall improvement in efficiency. In the Indian context, the financial sector, especially the banking sector, has been a major beneficiary from the inroads made by IT. Many new processes, products and services offered by banks and other financial intermediaries are now IT-centred. Effective integration of technology with sound business practices requires business process re-engineering and banks in India need to follow up on the beginnings made in this regard.

Newer delivery channels to customers—Automated Teller Machines (ATMs), networking of ATMs in the form of Shared Payment Networks, Internet Banking—and implementation of core banking solutions by most banks are some examples.

Reserve Bank of India (RBI) has played a proactive role in the implementation of IT in the banking sector. IT-based initiatives are focused on meeting the three-pronged objective of better house keeping, improved customer service and overall systemic efficiency.

## 8.1 Electronic Money (E-Money)

In recent years, there has been a gradual switchover from the use of paper-based payments media to those based on electronics. While the basic characteristics of these new instruments are by and large similar to those of paper-based instruments, the former present a different set of challenges to policy makers. E-money is one such new product which has appeared on Indian horizon recently.

**8.1.1 Definition and Kinds of E-Money:** E-money may be broadly defined as an electronic store of monetary value on a technical device used for making payments to undertakings other than the issuer without necessarily involving bank accounts in the transaction, but acting as a prepaid bearer instrument. E-money can be classified into the following two broad categories:

1. Pre-paid stored value card (also called electronic purse).
2. Pre-paid software-based product that uses computer networks such as internet (also referred to as digital cash or network money).

The stored value card scheme typically uses a microprocessor chip embedded in a plastic card while software based scheme typically uses specialised software installed in a personal computer. The stored value card could be of three types:

1. Single-purpose card.
2. Closed-system or limited-purpose card.
3. General-purpose or multi-purpose card.

The single-purpose card generally with a magnetic chip recording the amount of fund therein is designed to facilitate only one type of transaction, e.g. telephone calls, public transportation, laundry, parking facilities etc. In this case the distinguishing point is that the issuer and the service provider (acceptor) are identical for the cards. These cards are expected to substitute coins and currency notes.

Closed-system or limited-purpose card is generally used in a small number of well-identified points of sale within a well-identified location such as corporate/university campus.

Multi-purpose card can perform a variety of functions with several vendors, viz. credit card, debit card, stored value card, identification card, repository of personal medical information etc. The importance of these cards is especially underscored with respect to regulatory oversight, restrictions on issuers and their implications for monetary policy. These cards may reduce demand for current accounts in the bank for likely reduction in transaction costs, and prudent portfolio management.

It is important to distinguish here the so-called *access* products, e.g. credit card and debit card from e-money. The former typically require a telephone or a personal computer with appropriate software to access the customer account before transferring the value while under e-money, the amount of value is already embedded and it may be increased or reduced without necessarily involving a personal bank account. In a sense, e-money can be construed as an electronic form of traveller's cheques (TCs). In both cases, the user pays for the instrument upfront.

**8.1.2 Benefits of E-money:** Use and spread of e-money depends upon the incentives of issuers, merchants and consumers. For issuers, the principal motivation arises from revenues to be earned from investment of outstanding balances (i.e. float income), savings of costs from reduced handling of cash and incentive of offering fee-based service to consumers and merchants besides the larger security emanating from audit trail of transactions and improved management information system.

For merchants, the trade-off would be between reduction in cost of handling physical cash vis-à-vis cost of putting in place necessary infrastructure.

For consumers, the use of e-money would depend upon the perceived security and privacy of e-money and ease with which e-money could be used which would again depend upon willingness of merchants to accept e-money. In other words, apart from security, privacy and ease, network effects are also important for wider use of e-money.

An economy like India, where cash transactions are very high, could benefit from using e-money through cost savings from printing and minting of smaller denomination notes and coins and eliminating the cost of handling, storing, transporting and insuring currency. These should also improve operational efficiency of the financial sector as also extension of banking to the urban poor and rural communities besides facilitating E-governance initiatives of governments. However, such benefits should be weighed against the need to build up the costly infrastructure to operate nationwide cashless retail payment system.

**8.1.3 E-Money in India:** In India, where cash transactions are high in number, the use of E-money can be beneficial in terms of reduced miscellaneous costs, viz. cost of printing and minting of smaller denomination notes and coins and transportation and storage costs. However, certain additional costs for setting up of network infrastructure to operate nationwide are also associated with it. The RBI has been partnering a multi-application smart card project, under the aegis of the Ministry of Communications and Information Technology, Government of India, to run a pilot project on the use of multi-application smart cards in the country. Various issues relating to technology, security, regulatory and supervisory concerns and legal implications have been examined to make the use of smart cards a viable proposition after the conclusion of the pilot project. The project is aimed at combining applications relating to banking, insurance, postal services, identification, etc in a single card.

The Reserve Bank of India had constituted a Working Group on Electronic Money (Chairman: Zaire J. Camas) in January 2002 which submitted its Report in July 2002. The Group identified certain areas of concern from the point of view of the central bank in the context of more widespread use of e-money. Some of the suggestions made by the Group included: (a) multi-purpose e-money to be issued only by authorised banks on a credit basis which should be strictly regulated and closely monitored, (b) redeemability should be ensured in order to preserve the unit of account function of money as well as to control money supply in the economy, and (c) reporting of e-money statistics should be ensured for the purposes of monetary policy and protection against criminal abuse such as money laundering.

**8.2 Technological Solutions for Financial Services**

Most of the initiatives regarding technology are aimed at providing better and more efficient customer service by offering multiple options to the customer. The death of distance, which is a by-product of technology, has become a reality in the banking sector. Technology is also playing a key role in banks' strategy for gaining a competitive edge.

Many banks have commenced the process of setting up core banking solutions, which are at various stages of implementation. While such systems are already in place in respect of new private sector banks, a few old private sector banks and public sector banks are also quickly moving to set up such processes. This would benefit the customer in the banking sector. Computerisation of the business of banks has been receiving high importance. The public sector banks have already crossed the 70 percent level of computerisation of their business. The directive from the Central Vigilance Commission (CVC) to achieve 100 percent computerisation has resulted in renewed vigour towards computerisation of branches.

Networking has been receiving focused attention by banks. This activity is also being monitored by the RBI. Most banks have their own corporate networks to facilitate inter-branch and branch-controlling office communication in an electronic mode. Inter-bank and inter-city communication takes place through the Indian Financial Network (INFINET). As part of the INFINET, the terrestrial lines have been augmented to provide for increased data transfer capabilities. All these have resulted in the dependence of banks on network-based computing which has benefited the customer.

INFINET also provides for safe and secure transmission of electronic messages with the use of Public Key Infrastructure (PKI) which has the legal backing of Information Technology Act, 2000. It also provides for messages to flow in a structured environment, using the Structured Financial Messaging Solutions (SFMS), which provides for inter-operability of messages so that straight through processing (STP) is achieved.

Another major development witnessed in recent years is the growth in multiple delivery channels to customers such as internet-based banking, mobile banking and anywhere banking. This has benefited the customers and the banks alike. While customers have now a wide variety of options to choose from, banks have been able to reduce costs which have had a positive impact on their profitability.

These developments have, however, also posed certain challenges. In a world where geographical barriers are losing significance, it is essential that security is given prime importance in a trans-national scenario where large sums of money are at stake. While the challenges relating to physical security could be confronted with relative ease, the position is much more complicated in respect of IT security. The RBI has, therefore, provided guidelines on information system (IS) security as also IS audit which banks can use for their benefit. These are generic in nature and do not have any prescriptive tones.

**8.2.1 National Electronic Fund Transfer (NEFT):** The NEFT was launched by the RBI in November 2005 as a more secure, nation-wide retail electronic payment system to facilitate funds transfer by the bank customers, between the networked bank branches in the country. It has, however, been observed that the public sector banks are not the most active users of this product and the majority of NEFT outward transactions are originated by a few new-generation private sector banks and foreign banks.

In order to popularise the e-payments in the country, the RBI, on its part, waived the service charges to be levied on the member banks, till March 31, 2009, in respect of the Real Time Gross Settlement (RTGS) and NEFT transactions. The RBI also provides, free of charge, intra-day liquidity to the banks for the RTGS transactions. The service charges to be levied by banks from their customers for RTGS and NEFT have, however, been deregulated and left to the discretion of the individual banks. While some of the banks have rationalised their service charges and a few have made it even cost-free to the customers, there are also certain banks that have fixed unreasonably high service charges for providing these services to their customers.

**8.2.2 ATM Networks:** The National Financial Switch (NFS) network started its operations on August 27, 2004 and is owned and operated by Institute for Development and Research in Banking Technology (IDRBT), Hyderabad. NFS is one of the several shared ATM networks which inter-connect the ATM switches of the banks together and thus enable inter-operability of the ATM cards issued by any bank across the entire network. National Payments Corporation of India (NPCI) took over NFS operations from December 14, 2009.

While there are a few other ATM networks also functioning in the country, the NFS has emerged to be the largest one, with a network of 1,04,126 ATMs of 69 banks as in October 2012. The primary objective of any ATM network, like the NFS, is to make the ATM deployment more economical and viable for banks by pooling their respective ATM resources. The main advantage of an ATM network is that it obviates the need for having bank-specific multiple ATM installations in the same geographical area, thereby reducing the entailed costs for the banks but without compromising on the reach of the banks to their customers. From the perspective of customers, the ATM card of any bank can be used in any ATM—which enables more convenient and wider ATM access to the bank customers of varied banks.

It is noteworthy that the use of the ATMs connected through any of the ATM networks in the country became cost free for the customers from April 1, 2009. Thus, the networking of the ATMs across the country, by leveraging the technology, is indeed a very customer-friendly development.

According to RBI's *Report on Trend and Progress and Banking in India, 2011-2012* released in November 2012, the number of ATMs in the country stood at 98,074 out of which the percentage share of private sector banks stood at 38 percent, public sector banks at 33 percent, SBI and its associates at 27 percent and foreign banks at 2 percent.

**8.2.3 Credit Cards:** There has been phenomenal increase in the number of credit cards issued by the banks in India during the last few years. While the increasing usage of the credit cards is a welcome development in as much as it reduces reliance on currency for settlement of transactions, it also entails certain additional elements of operational risk.

In order to ensure orderly growth of the card segment of consumer credit and protect the interests of banks/non-banking financial companies (NBFCs) and their customers, the RBI constituted a Working Group on Regulatory Mechanism for Cards (Chairman: R. Gandhi). Based of the recommendations of the Working Group, draft guidelines on credit cards were framed by the RBI in June 2005.

The draft guidelines delineated the broad parameters that banks/NBFCs should, at the minimum, take into account with regard to the following:

1. Issue of cards with respect to clear mentioning of Most Important Terms and Conditions (MITCs).
2. Interest rates and other charges on customers.
3. Corrective mechanism on account of wrongful billing.
4. Use of direct sale agents and other agents for outsourcing various credit card operations.
5. Protection of customer rights especially in respect of right to privacy, customer confidentiality and fair practices in debt collection.
6. Redressal avenues of customer grievances.
7. Internal control and monitoring systems of the banks/NBFCs for such card operations.

The draft guidelines further stipulated that each

bank/NBFC must have a well-documented policy and a *fair practices code* for credit card operations. The *fair practices code* for credit card operations released by the IBA in March 2005 could also be used by banks/NBFCs. The bank/NBFCs code should, at the minimum, however, incorporate the relevant guidelines contained in the draft guidelines released by the RBI.

**8.2.4 Satellite Banking:** Having regard to much greater reliability of a satellite-based communication link for interconnecting the branches of the banks, particularly in the hilly areas and difficult terrain where terrestrial communication link is difficult to provide, the RBI had constituted a Technical Group to examine the proposal for providing satellite connectivity to the bank branches in such areas. The objective is to enable greater penetration of the electronic payment products in the hinterland areas also, by facilitating the integration of the rural and remote branches with the core banking solution platform of the banks and help them providing efficient banking services to their customers. Under the proposal, the RBI would be bearing a part of the leased rentals for the satellite connectivity, provided the banks use it for connecting their branches in the North Eastern States and in the under-banked districts in the rest of the country. Thus, the RBI would be providing an incentive to the banks for adopting the satellite communication technology for networking their branches in the remote areas. A discussion paper on this scheme was placed on RBI website in June 2008 for public comments.

## 8.3 Internet-banking and Electronic-banking

Electronic-banking (E-banking) is a generic term encompassing internet banking, telephone banking, mobile banking etc. In other words, it is a process of delivery of banking services and products through electronic channels such as satellite, telephone, internet, cell phone etc. The concept and scope of E-banking is still evolving. Internet-banking is a major component of e-banking.

Internet-banking offers different online services like balance enquiry, requests for cheque books, recording stop-payment instructions, balance transfer instructions, account opening and other forms of traditional banking services. These are mostly traditional services offered through Internet as a new delivery channel. Banks are also offering payment services on behalf of their customers who shop in different e-shops, e-malls etc.

**8.3.1 Internet Banking in India:** Internet banking—both as a medium of delivery of banking services and as a strategic tool for business development—has gained wide acceptance internationally and is fast catching up in India with more and more banks entering the fray. India can be said to be on the threshold of a major banking revolution with net banking having already been unveiled.

The growth potential of internet banking in India is immense. Further incentives provided by banks would dissuade customers from visiting physical branches, and thus get 'hooked' to the convenience of arm-chair banking. The facility of accessing their accounts from anywhere in the world by using a home computer with Internet connection, is particularly fascinating to non-resident Indians and high net-worth individuals having multiple bank accounts.

Costs of banking service through the Internet form a fraction of costs through conventional methods. The cost-conscious banks in the country have therefore actively considered use of internet as a channel for providing services. Fully computerized banks, with better management of their customer base are in a stronger position to cross-sell their products through this channel.

**8.3.2 Products and Services Offered:** Banks in India are at different stages of the web-enabled banking cycle. Initially, a bank, which is not having a web site, allows its customer to communicate with it through an e-mail address; communication is limited to a small number of branches and offices which have access to this e-mail account. As yet, many scheduled commercial banks in India are still in the first stage of Internet banking operations.

With gradual adoption of Information Technology, the bank puts up a web-site that provides general information on the banks, its location, services available e.g. loan and deposits products, application forms for downloading and e-mail option for enquiries and feedback. It is largely a marketing or advertising tool. Customers are required to fill in applications on the internet and can later receive loans or other products requested for at their local branch. A few banks provide the customer to enquire into his demat account (securities/shares) holding details, transaction details and status of instructions given by him.

Some of the banks permit customers to interact with them and transact electronically with them. Such services include request for opening of accounts, requisition for cheque books, stop payment of cheques, viewing and printing statements of accounts, movement of funds between accounts within the same bank, querying on status of requests, instructions for opening of letters of credit and bank guarantees etc.

Banks are thus looking to position themselves as one stop financial shops. Some banks have tied up with computer training companies, computer manufacturers, Internet Services Providers and portals for expanding their internet banking services, and widening their customer base. Setting up of Internet kiosks and permeation through the cable television route to widen customer base are other priority areas in the agendas of the more aggressive players.

Banks providing Internet banking services have been entering into agreements with their customers setting out the terms and conditions of the services. The terms and conditions include information on the access through user-id and secret password, minimum balance and charges, authority to the bank for carrying out transactions performed through the service, liability of the user and the bank, disclosure of personal information for statistical analysis and credit scoring also, non-transferability of the facility, notices and termination etc.

The race for market supremacy is compelling banks in India to adopt the latest technology on the Internet in a bid to capture new markets and customers. Under mobile banking services, customers can scan their accounts to seek balance and payments status or instruct banks to issue cheques, pay bills or deliver statements of accounts.

Compared to banks abroad, Indian banks offering online services still have a long way to go. For online banking to reach a critical mass, there has to be sufficient number of users and the sufficient infrastructure in place. Though various security options like line encryption, branch connection encryption, firewalls, digital certificates, automatic sign-offs, random pop-ups and disaster recovery sites are in place or are being looked at, there is as yet no Certification Authority in India offering Public Key Infrastructure which is absolutely necessary for online banking. The customer can only be assured of a secured conduit for its online activities if an authority certifying digital signatures is in place. The communication bandwidth presently available in India is also not enough to meet the needs of high priority services like online banking and trading. Banks offering online facilities need to have an effective disaster recovery plan along with comprehensive risk management measures. Banks offering online facilities also need to calculate their downtime losses, because even a few minutes of downtime in a week could mean substantial losses.

Some banks even today do not have uninterrupted power supply unit or systems to take care of prolonged power breakdown. Proper encryption of data and effective use of passwords are also matters that leave a lot to be desired. Systems and processes have to be put in place to ensure that errors do not take place.

Users of internet banking services are required to fill up the application forms online and send a copy of the same by mail or fax to the bank. A contractual agreement is entered into by the customer with the bank for using the Internet banking services. In this way, personal data in the applications forms is being held by the bank providing the service. The contract details are often one-sided, with the bank having the absolute discretion to amend or supplement any of the terms at any time. For these reasons domestic customers for whom other access points such as ATMs, tele-banking, personal contact, etc. are available, are often hesitant to use the Internet banking services offered by Indian banks. Internet Banking, as an additional delivery channel, may, therefore, be attractive/ appealing as a value added service to domestic customers. Non-resident Indians for whom it is expensive and time consuming to access their bank accounts maintained in India find net banking very convenient and useful.

**8.3.3 Challenges of Internet Banking:** Innovations are mixed blessings. They bring benefits as well as pose challenges. Innovations in information technology (IT) are no exception to this rule. The most prominent challenge arising from these innovations relates to the concept of security. Considering the scope for fraud in the electronic banking area and the possibility of contagion, the Reserve Bank of India—as a regulator and supervisor—has been proactive in addressing the risks associated with electronic banking. The RBI has been promptly addressing issues related to fraud with the use of electronic banking facility. Even after issuing guidelines for a secured electronic banking, the RBI advises the banks, from time to time, on control mechanisms to combat such frauds.

In India, the legal infrastructure for promoting internet banking has not yet been put in place in a comprehensive manner. India does not have a licensed certifying authority appointed by the Controller of Certifying Authorities to issue digital signature certificates. Also, India is not yet a signatory to the International Cyber Crime Treaty, which seeks to intensify co-operation among different signatory nations for exchanging information concerning crime and cyber criminals. Further, there are unresolved legislative issues related to cyber crime laws, clarification relating to regulatory authority over e-money products, consumer protection and privacy laws. To make the electronic banking operations in India more widespread, secure and efficient, these issues need to be addressed by relevant authorities.

As the banking practices and legislations concerning electronic banking are still in the process of evolution in India and abroad because of technological innovations, there is a need for a constant review of various legislations and regulatory framework relating to banking and commerce. The RBI is monitoring and reviewing the legal and other requirements of electronic banking on a continuous basis to ensure that the e-banking would develop on sound lines and the e-banking related challenges would not pose a threat to financial stability.

Government of India enacted The Information Technology Act, 2000, in order to provide legal recognition for transactions carried out by means of electronic data interchange and other means of electronic communication, commonly referred to as *electronic commerce*. The Act came into force with effect from October 17, 2000. The Act has also amended certain provisions of the Indian Penal Code, the Indian Evidence Act, 1872, The Bankers Book of Evidence Act, 1891 and Reserve Bank of India Act 1934 in order to facilitate e-commerce in India. However, this Act does not apply to:

1. A negotiable instrument as defined in Section 13 of the Negotiable Instruments Act, 1881.
2. A power-of-attorney as defined in Section 1A of the Power-of-Attorney Act, 1882.
3. A trust as defined in Section 3 of the Indian Trusts Act, 1882.
4. A will as defined in clause (*h*) of Section 2 of the Indian Succession Act, 1925.

5. Any contract for the sale or conveyance of immovable property or any interest in such property.
6. Any such class of documents or transactions as may be notified by the Central Government in the official Gazette.

Section 72 of the Information Technology Act, 2000 casts an obligation of confidentiality against disclosure of any electronic record, register, correspondence and information, except for certain purposes and violation of this provision is a criminal offence.

In the course of providing Internet banking services the banks in India are facing new challenges relating to online opening of accounts, authentication, secrecy of customers accounts, non-repudiation, liability standards and consumer protection etc.

**8.3.4 Internet Banking and the Reserve Bank of India:** The RBI has played a proactive role in the implementation of IT in the banking sector. IT-based initiatives are focused on meeting the three-pronged objective of better house keeping, improved customer service and overall efficiency. The RBI has come out with a Financial Sector Technology Vision Document outlining the approach to be followed for IT implementation for the medium-term period of about three years.

Several initiatives taken by the Government of India as well as the Reserve Bank of India (RBI) have facilitated the development of Internet-banking in India. As a regulator and supervisor, the RBI has made considerable progress in consolidating the existing payment and settlement systems, and in upgrading technology with a view to establishing an efficient, integrated and secure system functioning in a real-time environment, which has further helped the development of Internet-banking in India. The Government of India enacted the Information Technology Act, 2000 with effect from October 17, 2000, which provides legal recognition to electronic transactions and other means of electronic commerce.

The RBI has been gearing up to upgrading itself as a regulator and supervisor of the technologically dominated financial system. It issued guidelines on Risks and Control in Computer and Telecommunication System in February 1998 to all the banks advising them to evaluate the risks inherent in the systems and put in place adequate control mechanisms to address these risks, which can be broadly put under three heads, *viz.,* IT environment risks, IT operations risks and product risks.

The existing regulatory framework over banks has also been extended to internet banking. These guidelines cover various issues that would fall within the framework of technology, security standards and legal and regulatory issues. Virtual banks, which have no offices and function only online are not permitted to offer electronic banking services in India and that only banks licensed under the Banking Regulation Act and having a physical presence in India are allowed to offer such services. Further, banks are required to report to the RBI every breach or failure of security systems and procedures in Internet banking, while the RBI at its discretion may decide to commission special audit/inspection of such banks.

As per guidelines, banks no longer need any prior approval of the Reserve Bank for offering the internet banking services. Nevertheless, banks must have their internet policy and they need to ensure that it is in line with parameters as set by it.

RBI has taken the initiative for facilitating real time funds transfer through the Real Time Gross Settlement (RTGS) System. Under the RTGS system, transmission, processing and settlements of the instructions will be done on a continuous basis. Gross settlement in a real time mode eliminates credit and liquidity risks. Any member of the system will be able to access it through only one specified gateway in order to ensure rigorous access control measures at the user level. Further, generic architecture, both domestic and cross border, aimed at providing inter-connectivity across banks has been accepted for implementation by RBI.

With the process of dematerialisation of shares having gained considerable ground in recent years, banks have assumed the role of depository participants. In addition to customers' deposit accounts, they also maintain demat accounts of their clients. Online trading in equities is being allowed by SEBI. This is another area which banks are keen to get into.

Wireless Application Protocol (WAP) telephony is the merger of mobile telephony with the Internet. It offers two-way connectivity, unlike Mobile Banking where the customer communicates to a mailbox answering machine. Users may surf their accounts, download items and transact a wider range of options through the cell phone screen. WAP may provide the infrastructure for P2P (person to person) or P2M (person to merchant) payments. It would be ideal for transactions that do not need any cash backup, such as online investments. Use of this cutting edge technology could well determine which bank obtains the largest market share in electronic banking.

Keeping in view all the above developments, internet banking is likely to grow at a rapid pace and most banks will enter into this area soon. Rapid strides are already being made in banking technology in India and Internet banking is a manifestation of this. A sea change in banking services is on the cards. It would, however, be essential to have in place a proper regulatory, supervisory and legal framework, particularly as regards security of transactions over the internet.

## 8.4 Technological Infrastructure for Government Securities Market

As part of financial sector reforms, the RBI has taken several initiatives for developing the technological infrastructure for the efficient functioning of the government securities market. These measures were accompanied by an assessment of the risk management systems under the new institutional arrangements.

**8.4.1 Trading Infrastructure:** A well-developed government securities market requires a system of transparent pricing and allotment mechanism which reduces transaction cost and improves market efficiency. In June 1994, the National Stock Exchange (NSE) introduced a transparent fully-automated screen-based trading system known as National Exchange for Automated Trading (NEAT) in the wholesale debt market (WDM) segment for facilitating trading in various debt instruments, including government securities.

In order to facilitate easier access, wider reach and active participation in the government securities market, a facility of retail trading in stock exchanges, *viz.* National Stock Exchange (NSE), Bombay Stock Exchange (BSE) and Over the Counter Exchange of India (OTCEI) was provided from January 16, 2003. Primary (urban) co-operative banks and FIs were permitted to transact in dated Central Government securities—in dematerialised form on automated order-driven systems of stock exchanges—from March 13, 2003 and from June 1, 2003, respectively.

**8.4.2 Negotiated Dealing System (NDS):** The Negotiated Dealing System (NDS) was operationalised from February 15, 2002 to provide, *inter alia*, an on-line electronic bidding platform for primary auctions in Central/State Government securities and OMO/LAF auctions.

The RBI introduced the NDS with the objectives of: (a) ushering in an automated electronic reporting and settlement process, (b) facilitating online electronic bidding in primary auctions and (c) providing an electronic dealing platform for trading in government securities in the secondary market. The NDS, which is available on a secure network, i.e. Indian Financial Network (INFINET) to a closed user group, facilitates straight-through settlement of secondary market transactions, thereby enhancing transparency and transactional efficiency. The NDS has greatly enhanced operational efficiency of the market by automating the flow of traded data into the settlement system. It has also facilitated dissemination of price information almost on a real time basis to market participants, enabling them to execute trading decision more effectively. The NDS has, however, gained popularity more as a reporting platform for the trade concluded bilaterally in OTC markets than as a trading platform as originally envisaged.

In order to provide the NDS members with a more efficient trading platform, the NDS-Order Matching (NDS-OM) trading module was operationalised on August 1, 2005 on the basis of the recommendations of the Working Group on Screen-based Trading in Government Securities (Chairman: R.H. Patil).

The NDS-OM is an anonymous order matching system which allows straight-through processing (STP). It is purely order-driven with all the orders matched on a strict price/time priority basis. The executed trades flow straight to the Clearing Corporation of India Ltd. (CCIL) in a ready-for-settlement stage. The CCIL is the central counterparty to each trade undertaken on the system. Participants have the option of using the NDS or the NDS-OM for their trading operations. The settlement of both types of transactions is, however, integrated. In the first phase of operationalisation of NDS-OM, only RBI regulated entities, i.e. banks, PDs and FIs were permitted to access the system. Subsequently, insurance companies were also allowed access. Those insurance companies, which did not have a current account with the RBI, were allowed to open a special current account with it.

Consequent to the announcement made in the Union Budget for 2006-07, access to NDS-OM was further extended to all qualified mutual funds, provident funds and pension funds. While large participants in these categories can have a direct access to NDS-OM system by obtaining the direct membership, small participants are envisaged to access the system through their principal member (CSGL route). The NDS-OM system has been well received by market participants as it enhances operational and transactional efficiencies. This system has provided an efficient price discovery mechanism, reducing intra-day price volatility.

**8.4.3 Clearing Corporation of India Ltd. (CCIL):** CCIL was established on February 15, 2002 to act as the clearing house and as a central counterparty for transactions in government securities. The CCIL has 154 members participating in the securities settlement segment. The establishment of CCIL has ensured guaranteed settlement of trades in government securities, thereby imparting considerable stability to the markets. Through the multilateral netting arrangement, this mechanism has reduced funding requirements from gross to net basis, thereby reducing liquidity risk and greatly mitigating counterparty credit risk. The CCIL has been equipped with the risk management system to limit the settlement risk. Operational guidelines were issued to the CCIL in April 2003 for a limited purpose government securities lending scheme. Accordingly, the CCIL has been permitted to enter into an arrangement with any of its members for borrowing government securities for the purpose of handling securities shortage in settlement. All transactions in government securities concluded or reported on NDS as well as transactions on the NDS-OM have to be necessarily settled through the CCIL. The net obligations of members are arrived at by the CCIL for both funds and securities and then sent to the RBI for settlement under the DvP mechanism.

As a step towards introducing the national settlement system (NSS) with the aim of settling centrally the clearing positions of various clearing houses, the integration of the integrated accounting system (IAS) with the real time gross settlement system (RTGS) was initiated in August, 2006. This facilitates settlement of various CCIL-operated clearings (inter-bank government securities, inter-bank foreign exchange, CBLO and National Financial Switch) through multilateral net settlement batch (MNSB) mode in the RTGS in Mumbai.

**8.4.4 Settlement Cycle:** The government securities market earlier followed both T+0 and T+1 settlement systems. In order to provide participants with more processing time and facilitate better funds and risk management, the settlement cycle for secondary market government securities transactions was standardised to T+1, effective May 11, 2005.

**8.4.5 When Issued (WI) Market:** As part of the restructuring of the debt issuance framework, guidelines for trading in WI market [1] were issued by the RBI in May 2006. A limit of 5 percent (only buy side) of notified amount was prescribed for banks and 10 percent (both buy and sell) for PDs. WI trading, which commenced in August 2006, was initially permitted in reissued securities. It takes place from the date of announcement of auction till one day prior to allotment of auctioned securities. The revised guidelines extending WI trading to new issuances of Central Government securities on a selective basis were issued in November 2006.

## 8.5 Computerisation in Regional Rural Banks (RRBs)

The need for computerisation in RRBs had been felt since long and some RRBs had been taking some steps in that direction. In July 2001, Government of India and NABARD advised the RRBs to initiate immediate steps so that Head Office, Area Office and a minimum of 50 percent of the branches were computerised in a phased manner in the next 5 years. Sponsor banks were also advised to formulate RRB-wise Action Plans, keeping in view the financial position of RRBs, infrastructure facilities available in their command area and the business potential of the RRB branches. Necessary support to implement the programme was also required to come from the sponsor bank.

NABARD made a beginning by extending support to select RRBs by providing PCs, peripherals, standard software packages as also customised MIS package and training inputs under its Swiss Agency for Development and Cooperation (SDC) programme. A review made by NABARD in respect of 152 RRBs in 2005 indicated that only 9 banks had reported achieving 100 percent computerisation. Most of the RRBs were lagging behind the target of computerising 50 percent of their branches.

The Task Force on Empowering RRBs for Operational Efficiency (Chairman: K.G. Karmakar) had made the following observations regarding the computerisation scenario in RRBs:

1. Level of computerisation is far behind schedule.
2. Even RRBs of same sponsor banks although following the same technology, are at different levels of computerisation due to varied financial health, environment, local conditions which has become a problem when such RRBs have been amalgamated.
3. Computerisation needs both capital and recurring expenses which are beyond the capability of some of the banks.
4. RRBs, which have progressed well, have adopted a particular technology mainly of the sponsor bank and invested sizeable amounts.
5. A few RRBs reaped the benefit of hosting their own website by way of mobilisation of deposits from non traditional segments.
6. A few of them established direct links with money market operators for investments.
7. Internet facilities have been used for faster communication with techno savvy customers and other agencies.

These observations highlighted the varying status of RRBs with regard to computerisation.

The status of computerisation at the end of March 2007 as reported by 76 of the 96 RRBs to the Committee to Formulate a Comprehensive Human Resource Policy for RRBs (Chairman: Y.S.P. Thorat) indicated that 35 of the reporting banks had achieved 100 per cent computerisation while 25 banks had achieved 50 to 100 per cent computerisation.

**8.5.1 Working Group on Technology Upgradation of Regional Rural Banks, August 2008:** In RBI's Mid-term Review of Annual Policy for 2007-08, it was stated that in order to prepare RRBs to adopt appropriate technology and migrate to Core Banking Solutions (CBSs) for better customer services, it was proposed to constitute a Working Group with representatives from RBI, NABARD, sponsor banks and RRBs for preparing a road-map for migration to core banking solutions by RRBs.

Accordingly, a Working Group (Chairman: G. Srinivasan) was set up with the following terms of reference:

1. To determine the nature of core banking solutions required for RRBs having regard to the range of their business and customer service;
2. To examine various options including extending core banking solutions of the sponsor banks to their RRBs with necessary modifications and firewalls;
3. To estimate the likely costs involved and funding and training arrangements that may be necessary; and
4. To draw out a time-bound road map for implementation.

The Working Group was also requested to examine the possibility of using solar power generating devices for meeting the power requirements of RRBs, especially in remote areas.

The Working Group was of the opinion that RRBs cannot afford to remain isolated from the technological developments sweeping the banking sector. With the commercial banks racing towards a higher degree of technological sophistication, the RRBs would be required to adopt technology for improving the quality of their customer service. However, given the different levels at which the different RRBs are presently placed in regard to their status of computerisation, a "one strategy fits all" approach may not be workable.

The Working Group was of the view that as a matter of policy, all RRBs should begin moving towards CBS. The CBS in RRBs should be geared towards better management control and monitoring, wider range of services offered and enhanced level of customer satisfaction. Adoption of CBS would lead to uniformity in work environment, more informed decision making, centralised processing and better MIS and reporting and improved regulatory compliance.

The Working Group also examined the use of solar power devices such as solar photovoltaic cells for RRBs, particularly in respect of branches located in remote areas which are not assured of continuous power supply. In this context, the experience of one of the RRBs in Uttar Pradesh, which has made a beginning in this direction, was also studied.

Funding the cost of moving towards CBS is an important issue for RRBs, particularly as several RRBs continue to incur operating losses. Further, many RRBs still continue to have accumulated losses and even negative net worth. Despite this position, there are RRBs which are healthy and making sustained profits. The Working Group considered this aspect and was of the view that funding support may be provided to all RRBs in the matter of adopting CBS, irrespective of their financial position. This would enable the entire sector to move uniformly towards adopting CBS.

A summary of the recommendations of the Working Group is given below.

1. The significant differences in the level of computerisation between different RRBs need to be taken into account while preparing any plan for further technology upgradation of RRBs.
2. As a matter of policy, all RRBs should begin moving towards CBS. The CBS in RRBs should be geared towards better management control and monitoring, wider range of services offered and enhanced level of customer satisfaction. Adoption of CBS would lead to uniformity in work environment, more informed decision making, centralised processing and better MIS and reporting and improved regulatory compliance.
3. The respective sponsor banks may be given the option of selecting the service provider for their RRBs. This provider may be same as for the sponsor bank or different as per the requirement of the RRBs. In case a sponsor bank has developed its own software, it may consider making the same available to its RRBs.

4. Any technology platform for CBS needs to be reliable, scalable, easily available, manageable and secure. For this, the background, credentials and experience of the vendor are extremely important. It would therefore be prudent to rely on the trusted and established vendors to provide services to RRBs.
5. Sponsor banks must own major responsibility in taking RRBs on the path to CBS. However, the requirements of the RRBs have to be borne in mind as these banks may not require the full suite of services that are available in sponsor banks and hence a plain vanilla tailored to their needs should be provided.
6. RRB branches in remote areas where power is a serious problem may take recourse to powering their branches through solar power in a CBS scenario.
7. Funding support may be provided to all RRBs in the matter of adopting CBS, irrespective of their financial position. This would enable the entire sector to move uniformly towards adopting CBS.
8. While sponsor banks may contribute 25 percent of the cost required for taking their RRBs to CBS, the remaining cost may be borne by Reserve Bank through Institute for Development and Research in Banking Technology (IDRBT). The IDRBT, in turn would provide necessary funds to the sponsor banks for their sponsored RRBs. The support could be in the form of interest free loans, repayable in three years.
9. The training requirements of staff of RRBs on account of moving towards CBS would have to be met by sponsor banks through their training establishments. The sponsor banks should also meet the costs incurred in this regard.
10. The roll-out of the road map for implementation of CBS in RRBs should start at the earliest. RRBs in which 100 per cent branches are computerised, should implement CBS in all their branches by September 2009.
11. The remaining RRBs may also start the process of moving

their branches to CBS at the earliest. Such RRBs should implement CBS in at least 25 percent of their branches by September 2009 and at least 50 percent of their branches by September 2010.

The remaining branches may be endeavoured to be covered by September 2011. By that date, at least 90 percent of the business of the banks should be on CBS.

12. The responsibility of ensuring that the time frame suggested is adhered to would be that of the sponsor bank.
13. All new branches opened after September 2009 may be made CBS compliant from day 1.

**8.6 Technological Upgradation of Stock Exchanges**

There are 23 recognised stock exchanges in India, including the Over the Counter Exchange of India (OTCEI) for small and new companies, the old established Bombay Stock Exchange (BSE) and the newer National Stock Exchange (NSE) that was set up as a model exchange to provide nation-wide services to investors.

The Bombay Stock Exchange (BSE) was set up in 1875 as 'The Native Share and Stock Brokers Association'. Bombay Stock Exchange is the largest of 25 exchanges in India.

Equity trading is most active in the two major competing stock exchanges, viz. NSE and the BSE.

Trading infrastructure in the stock exchanges is anonymous and order-driven, with all orders from market participants being matched based on strike price/time priority.

As at end-April 2012, 5,133 companies were listed on BSE and over 9,275 scrips were being traded.

The Indian capital market is significant in terms of its degree of development, its volume of trading and of the quality of its automated trading and settlement. The capital markets—though they have suffered from periodic scandals and problems in the past—look much closer in quality to those of developed markets than those in the majority of emerging markets. The range and quality of available securities is evidenced by the substantial flows of foreign institutional investors.

The competition between NSE and BSE is a unique one by international standards, where both exchanges are in the same city and have the same trading hours. All major stocks trade on both exchanges, so the exchanges compete for order flow, and not just listings. The rise of NSE has proved to be a powerful spur to reforms at the BSE. Months after NSE started operations, the BSE also launched electronic trading, and improved rules governing admission of corporate and foreign brokerage firms. Presently, the BSE also uses an open electronic limit order book market, using satellite communications to reach locations outside Bombay.

BSE—one of the oldest exchanges in the world—accounts for the largest number of listed companies and has also started a screen-based trading system with the introduction of the Bombay On-Line Trading system. NSE—which in the recent past has accounted for the largest trading volumes—has a fully automated screen-based system that operates in the wholesale debt market segment as well as the capital market segment. BSE and NSE account for the large majority of trading volumes owing to establishment of screen-based trading and terminals in all major and many minor cities. Presently, small regional exchanges are virtually moribund.

Associated with the NSE are the National Securities Depositary Ltd. (NSDL) and the National Securities Clearing Corporation Ltd (NSCCL). The NSDL acts as the registrar for what are now predominantly dematerialized securities and the NSCCL as a clearing house.

The setting up of the NSE as an electronic trading platform set a benchmark of operating efficiency for other stock exchanges in the country. The establishment of NSDL in 1996 and Central Depository Services (India) Ltd. (CSDL) in 1999 has enabled paperless trading in the exchanges. This has also facilitated instantaneous electronic transfer of securities and eliminated the risks to the investors arising from bad deliveries in the market, delays in share transfer, fake and forged shares and loss of scrips. The electronic fund transfer (EFT) facility combined with dematerialisation of shares has created a conducive environment for reducing the settlement cycle in stock markets.

**8.6.1 Corporatisation and Demutualisation of Stock Exchanges:** Stock exchanges all over the world have been traditionally formed as *mutual* organisations. The trading members not only provide broking services, but also own, control and manage such exchanges for their mutual benefit. In India, NSE was set up as a demutualised corporate body, where ownership, management and trading rights are in the hands of three different sets of groups from its inception. The Stock Exchange, Mumbai—one of the two premier exchanges in the country—has since been corporatised and demutualised and renamed as the Bombay Stock Exchange Ltd. (BSE).

Corporate governance has emerged as an important tool for protection of shareholders. The corporate governance framework in India has evolved over a period of time since the setting up of the Kumar Mangalam Birla Committee by SEBI. According to the Economic Intelligence Unit Survey of corporate governance across the countries (2003), India was rated the third best country to have good corporate governance code after Singapore and Hong Kong. India has a reasonably well-designed regulatory framework for the issuance and trading of securities, and disclosures by the issuers with strong focus on corporate governance standards.

To enhance the level of continuous disclosure by the listed companies, SEBI amended the listing agreement to incorporate segment reporting, related party disclosures, consolidated financial results and consolidated financial statements. The listing agreement between the stock exchanges and the companies has been strengthened from time to time to enhance corporate governance standards.

# 9

# Problems, Policy Implications and Prospects of ICT Sector

## 9.1 Problems of ICT Sector

The ICT sector is one of the most important and fast growing sectors in the world. ICT plays a vital role in the national and international economy. A number of favourable factors have helped in the rise of ICT industry in India. It include the availability of a large pool of talented English speaking manpower at low wages, a tradition of logic and mathematics, past investment by government in national innovation system, availability of infrastructure and communication links, favourable policy regime and growing scarcity of trained manpower in the West. However, it will be a challenge to sustain the growth rates achieved in the past in the medium term and to fulfil the NASSCOM-McKinsey-NIITSD targets of IT industry growth.

**9.1.1 Shortage in Supply of Suitable Talent:** A major concern for the IT industry is its inability to meet emerging global demands due to lack of sufficient number of skilled workers. With the improvement in the quality and type of works (high-end) being handled in India, the firms are finding it difficult to find appropriate workers with required skill base and retain them.

To achieve US$ 86 billion in exports revenue by 2012 and to continue to grow the domestic IT industry, the IT software and service industries will need to employ nearly 3.7 million professionals. The industry employed 2.1 million professionals in 2008 (NASSCOM, 2008). Therefore, the higher education system needs to produce enough graduates to provide another 1.6 million suitable professionals by 2012 (Government of India, 2006a, 2007a). There is absolute dearth of specialists with critical skills in science and technology and related fields of study. For instance, the graduates in medical and veterinary sciences form the smallest fraction of India's S&T manpower. Engineering degree holders are in short supply.

The IT and related sectors are expected to throw up about 8 million new jobs in the near future, but engineering degree holders constitute just 12.67 percent of available stock of S&T manpower (Chadha, 2008).

The NASSCOM-McKinsey study team has analyzed three types of skills, which are needed by the industry. These include technical skills, functional skills and life skills.

The majority of educational institutes provide technical skills. Some institutes may be able to provide functional skills because of excellent infrastructure and few are able to impart life skills because of quality of inputs and environment.

According to NASCCOM-McKinsey study, the technical skills required in 1998 was ¾ the (76 percent) of the total manpower and it would be reduced to half in 2008 whereas the requirement for functional and web skills would double.

This would mean that work force development through non-formal sector needs to play important role (Government of India, 2002b).

In order to cope with shortage of skilled workers, efforts are being made by individual firms in the industry and the industry association NASSCOM.

For example, NASSCOM has initiated NASSCOM Assessment of Competencies (NAC) program for the potential employees in the BPO industry.

The NAC scheme aims to create a talent pool within the BPO industry. Moreover, many large companies spend 6-10 percent of their revenue on the training program and have started state-of-the-art training centres which can train a large number of students/employees at any given time.

**9.1.2 Quality of Education:** India spends almost the same on education as many developing nations. Also the percentage of students studying in the science stream is also equal, but the number of people doing research in India is far less. This shows that country does not provide quality education that motivates students to undertake research. The number of persons doing research and development in the U.S.A is 4605 per million, in Japan it is 5287 per million and in China, 1564 per million. In India, there are only 119 researcher per million of population. This is an area of major concern and worry (Sharma and Singla, 2009).

A large proportion of educated manpower ranging between 50-60 percent has remained stuck at the undergraduate level, both in science and non-science discipline. They are mainly trained in basic sciences and remain largely unemployable and represent the largest chunk of educated manpower.

The rising stock of scientifically groomed and technically trained manpower, notwithstanding, the quality of the technical talent of our manpower has been found to be below par in terms of matching the requirement of the competitive world. American and European MNCs which had earlier been outsourcing jobs recklessly to the Indian job market have now started finding the capabilities and talent of Indian engineering specialists to be inadequate. They have started cutting jobs and pay pockets (Chadha, 2006, 2008).

**9.1.3 Diminishing Labour Cost Advantages:** With the growing scarcity of skilled professionals, salaries in the ICT sector have been rising. Wages and salaries in the ICT sector have risen by 19.17 percent per annum during the period 1991-07. Rising salaries have reduced the margins of advantage in wage cost that Indian ICT companies had enjoyed in the initial period of growth. The strategy of competing based on cost will become increasingly difficult for Indian companies especially with emerging competition from other low wage countries.

**9.1.4 High Attrition Rates:** The IT enabled services (BPO) industry is being looked upon as the next big employment generator. However, it is not an easy task to bridge the ever increasing demand and supply gap of professionals in this sector. Companies invest significantly in training entry-level employees. In India, the average attrition rate in the BPO sector is approximately 30-35 percent. It is true that this is far less than the prevalent attrition rate in the US market (around 70 percent), but the challenge continues to be greater considering the recent growth of the industry in the country. Keeping low attrition levels is a major challenge as the demand outstrips the supply of good professionals by a big margin. High attrition levels stem from factors like high stress and lack of growth opportunities. The Indian BPO industry is characterized by ungodly work hours, never ending night shifts, high work targets and repetitive nature of the work, irate customers and insufficient holidays. Many of the workers suffer from identity crisis as they speak and work with the American name.

The high percentage of females in the workforce (constituting 30-35 percent of the total), adds to the high attrition rate. Most women leave their job either after marriage or because of social pressures caused by irregular working hours in the industry. All this translates into huge losses for the company, which invests a lot of money in training them. Many experts are of believe that all these challenges can turn out to be a real dampener in the growth of this industry. At least 60,000 of the 1.71 lakh workforce change jobs every year. About 80 percent of them look for better leaders. Team leaders want to upgrade to supervisors, quality professionals or operations heads. This means costs will rise making it difficult for small venture capital funded companies to survive (NASSCOM, 2008).

**9.1.5 High Rates of Emigration:** The number of Indian students going abroad annually is over one lakh and is projected to cross two lakh by the end of this decade. In comparison, the number of foreign national students coming for studies in India is estimated at less than 10,000 each year. As a result, the net annual outflow on account of education is estimated to be US$ 3.50 billion and is likely to continue to rise rapidly (Government of India, 2006).

**9.1.6 Fragmented Policy Regime and Inadequate Infrastructure:** Policy for development of ICT sector is carried independently by two agencies of the Government namely-Department of Information Technology and Department of Telecommunication. Telecom infrastructure plays the crucial role in the growth of the ICT industry. Therefore, unless planning for the Information and Communication Technology (ICT) is not carried under one roof, the desired results may not be achievable. Moreover, lack of coordination among Ministries/Departments/Organizations for other infrastructure like roads, rails, gas, oil pipes etc. have led to slower development of infrastructure in India.

Infrastructure that is of specific interest to the IT industry is the provision of telecommunication network. Software companies in India encountered a number of telecommunication problems including telecommunication links, poor transmission and high cost of installation and use. Such limitations in telecommunication facilities have discouraged foreign clients from considering offshore software development as a feasible option (Government of India, 2002b).

Indian high tech industries are facing problem of highways, modern bridges, pot-holed roads, excellent airports, reliable power, and clean water are in desperately short supply. India's overstressed power grid is one of the most obvious signs of lagging infrastructure development. It is estimated that between increased consumer demand and new industrial projects, over a hundred gig watts of new power capacity are needed by 2012. If our infrastructure is delayed, our economic development, job creation, and foreign investment are delayed.

**9.1.7 Data Security:** It has been pointed out that quite often the companies abroad are reluctant and sceptical about outsourcing high-end services due to varied reasons such as data security, quality and professionalism in remote locations and fear of losing control over core processes. Sharing of data to firms located abroad is often considered risky by many potential service providers. For instance, the growing protest in the U.S.A highlighting the possible adverse consequences of off-shoring led to certain legislations and changes in the policy framework (e.g. restriction on government department on off-shoring). Despite the wide recognition regarding higher and improving quality of services being provided by Indian firms, still there are discomforts for clients from the developed nations (Ramesh, 2009).

**9.1.8 Low Research and Development Thrust:** The Indian ICT sector has demonstrated little thrust and enthusiasm on research and development as compared with the knowledge intensity and its international orientation. The largest source for R&D in the country is the Department of Information Technology (DIT). The R&D expenditure spiralled up from ₹ 113.3 crore in 2002-03 to ₹ 245.22 crore in 2008-09 (Government of India, 2007, 2010). Research and development in ICT is of critical importance for the nation's security and fulfilling various societal objectives. Unfortunately, India's investment of 0.8 percent of GDP, in this crucial area is very low compared to most other countries.

Although leading companies such as TCS, Infosys and Wipro do have R&D labs, the proportion of revenue devoted to it is rather low compared with IT companies in the developed world. In addition to it, the industry's recent problems include rapid appreciation of rupee against dollar, recession in the US (the Indian ICT industry's main market), the uncertain disposition of the global economy, an unclear tax holiday structure and a volatile stock market. The rupee's appreciation against dollar, almost 12 percent in 2007 has put pressure on profit margins of IT industry (NASSCOM, 2008).

## 9.2 Policy Implications

Indian ICT sector has attained a unique status the world over and has been recognized as key for the future development. To reap the full benefits of ICTs, India needs to boost the use of ICTs throughout the country, to address the local and national dimensions of ICT infrastructure and connectivity (access, content and language). Some critical policy implications may include the following:

1. Improve efficiency in the use of public resources in the education system and make the education system as a whole more responsive to market needs. The Government should ensure the expanded access to education that fosters critical thinking and learning skills for all. To strengthen the knowledge economy of India, quality of primary and secondary education should be enhanced with special emphasis on issues related to teacher vacancies and absenteeism, reversing high drop-out rates and correcting inadequate teaching and learning materials. This is especially important for India to meet the goal of providing eight years of schooling for all children by 2010 (Dahlman and Utz, 2005). The quality of all higher educational institutions, not just a few excellent one (such as the IITs) should be raised. To improve the operating environment for higher education, a shift in the role of government from managing the administrative aspects of the institutions to becoming an architect of education standard is required. It will improve the quality of academic programs and procedures ensure equity and coordinate a system with multiple players and multiple pathways to learning.
2. The curriculum of tertiary education should be reformed to include skill and competencies for the knowledge economy (communication skills, problem-solving skills, creativity and teamwork). Syllabus in computer engineering, electronics and IT in various technical institutions should be updated to meet the demands of industry. It may not be possible for the academic institutions to impart all the skills essential for a quality professional. A mechanism needs to be devised to bridge the gap between what is offered by academic institutions and what is required by the industry. This can be done

through bridge course in the skill formation as well as for high end applications and product development activities. Bridge program can increase the IT workforce by training graduates, non-computer science engineers and computer science engineers (Government of India, 2002b).

3. There is need to set up a Core Group/Committee from industry and educational institutions to assess the future skills requirement and suitably design educational and training program. This program would identify current and emerging global markets and accordingly assess the needs for IT personnel in different sectors. The Core Group will also assess IT work force requirement on regular basis.
4. Given the varying standards of different institutes, a common nation-wide test for graduates is also necessary. The test must include the requirement of the ICT industry to make it useful for recruitments. It will also ensure that colleges or training institutes include these specific elements in their curriculum. A series of such benchmarks need to be created to certify candidates suitable for different levels of jobs, beginning with the entry level. Eventually, it should evolve into something along the lines of chartered accountants exam (Government of India, 2006).
5. The government should relax the bureaucratic hurdles and put in place better accreditation system for private educational institutions. To have international educational standards, partnerships between Indian and foreign universities should be established. It will attract and retain high quality staff and provide opportunities for students. The intake of students into science and engineering should be increased. This may require adding colleges and universities such as IITs and others modelled after them.
6. To make higher education more demand oriented, funding should be changed from an institution based approach (direct funding of colleges by state) to a more individual-based approach (e.g., state funded loans, scholarships to

students etc.). This can be facilitated through intermediaries such as banks or the setting up of educational credit institutions on the lines of existing agencies e.g. Federal students Aid (FSA) of the US. A large-scale merit-cum-means scholarship system can also be introduced to ensure equal opportunities in higher education irrespective of economic status (NASSCOM-McKinsey, 2005). A shift to demand-based funding will force colleges to innovate in courses offered and methods to satisfy market needs.

7. The State Governments can initiate discussions with financial institutions to provide loans at low interest rates to citizens for purchase of computers. Taxes may also be reduced on ICT hardware to bring down the cost of computers in the different states.
8. For the successful applications of ICT e.g. in e-governance among different Indian states, there should be scaling up of ICT initiatives. It will also bring the benefits of connectivity to rural community across the country. Partnerships among government agencies, research and academic institutions, private companies and non-governmental organizations (NGOs) should be strengthened to ramp up ICT infrastructure and achieve faster penetrations of ICTs (Gayathri and Vijaybaskar, 2003).
9. The intercity network and public transport infrastructure should be strengthened to decongest existing IT hubs. The industry should be decentralized beyond existing hubs by developing new townships. Action will need to be taken across various fronts including: master plan development, financial arrangements, land acquisition and auction, proper re-settlement of displaced persons, aviation and transportation planning and educational linkage.
10. Unlike old technologies, the new technologies are supply-driven and leave greater scope for the diffusion agents (non-governmental organization, government, private

sector and others) to influence diffusion process. Innovations like internet cafe, community internet centres, fixed lines and mobile phones, smart cards and satellite television can greatly offset the limits imposed by lower connectivity and poor information infrastructure. It will help reduce the digital divide across the country and improve the delivery of services to rural population.

11. The telecom service sector is the fastest growing sector of the Indian economy and it contributes significantly to GDP. However, the telecom equipment sector is not able to meet demand created by the telecom service sector. In 2006-07, 65 percent of total consumption of telecom equipment was met through imports (Sharma and Singla, 2009). Therefore, there is need to motivate growth of telecom equipment industry. This will also help the growth of other industries through forward and backward linkage.
12. The state governments should revise the ICT policies from time to time, e.g., the ICT policy of Pondicherry, Uttar Pradesh and Kerala were last updated in 1999, 2001 and 2004 respectively. The ICT policy of many State Governments does not have a section on security and legal policy framework, e.g., as in states like Chhattisgarh, Pondicherry (Government of India, 2007b). The Centre/State Government should address the issues of Intellectual Property Rights (IPR) in the ICT policy and effective legal machinery should be put in place to prevent piracy of ICT products. Government should establish world class centres of cyber-crime and dedicated courts for fast track resolution of IP related disputes in the country.
13. Government should have a clear policy defining the IPR and royalty sharing between the developing institutions, inventors and the industry. This will reduce the delays in taking up decision on transfer of technology and motivate the researchers to take up industry oriented R&D (Government of India, 2007a). In addition, direct rewards should be given to researchers and innovators who

generate patentable intellectual property.

14. ICT skills are linked almost to literacy. Devices that do not require literacy such as based on sound, touch, images and symbols are not widespread. Therefore, we require research to invent them.
15. The US continues to be the biggest market for the IT industry, accounting for over 60 percent of export revenue followed by Europe, which accounts for 20 percent. The Western markets are being the worst affected by the economic turmoil. Therefore, Indian IT industry needs to balance its geographical footprint to get a larger share of revenue from other parts of world like West Asia, Latin America, South Africa, Eastern Europe and Australia. Diversifying into new markets will not only counter the impact of recession in a particular geography, will also be a natural hedge against currency fluctuations. The recent moves under Obama Government to curb H1B visas (under which the US companies hire skilled foreign workers) and the stringent noise against off-shoring are now other reasons for the Indian IT industry to spread its wings to new geographies. However, the cost of acquisition is much higher and the margins much lower, but this investment needs to be done for the long-term.
16. Largely, the Indian IT firms have ignored the domestic market. Less than 20 percent of Indian population has access to a computer. With Indian economy, growing at around 8 percent and increasing IT spend; this is the time for the industry to look inwards. Hence, there is huge opportunity for Indian IT industry to grow within the country (Mohan, 2009).
17. The computer and the peripheral market in India are growing at a rapid pace. However, a poor investment climate, a policy structure non-conducive to manufacturing has failed to attract any significant investment in IT hardware segment. It is essential for government to give thrust to domestic consumption so that it provides an

incentive to IT industry to scale up their manufacturing operations. It will also lead to significant employment opportunities, not only direct but also indirect including those in research and new product development.

18. Government should ensure that small and medium enterprises (SMEs) can get the benefits offered under STP/SEZ scheme irrespective of their location. It should also encourage the use of SME/domestically developed IT in e-governance projects. Large companies (e.g., with a turnover of over ₹ 100 crore) may be offered tax benefits on costs incurred on the goods and services purchased from domestic SMEs through a subcontract model. This will boost SME growth in the sector (Government of India, 2006). Government may allow tax deduction of up to 20 percent of taxable income to all professionals working with SME companies for a minimum period of two years.
19. Indian IT firms also need to strengthen their global delivery models, which typically tend to be largely India-centric. Large clients are spreading their business across the world and expect the suppliers to service them in these locations too. At present, no single Indian company can compete with firms like IBM and Accenture when it comes to their global footprint. Therefore, it is imperative for Indian vendors to go beyond locations in India and to invest in technology and processes so that they can execute projects concurrently from multiple centres.
20. To nurture research and development activities, special Research and Education Zones (REdZ) should be established that carry excellent research and can attract top global talent. Research within each zone can be designed to focus on complementary areas fostering development of an innovation cluster. Ideally, REdZ should be located in the new townships and institutions should be given full freedom to operate. Market forces will determine their success or failure.

21. Another area requiring special attention is corporate governance. While the Satyam scam may be seen as an isolated incident, it has undoubtedly questioned the corporate governance practices of every Indian vendor. Customers will now be asking a lot more questions and vendors will need to be more proactive with disclosures.
22. Government should work towards 'free trade in services' agreement with UK and US. Department of Commerce may take that matter in Bilateral Trade Agreement for software industry with countries such as France, Germany, Japan, US and UK for taking full commitment in Mode 4 (presence of natural person) for market access, national treatment and issue of visas (Government of India, 2006).

**9.3 Prospects for ICT Sector**

In the current knowledge driven era, India has already developed a profile for the knowledge society. Broader than a knowledge economy, a knowledge society includes spiritual, social, intellectual and philosophical knowledge. In 2002, India's President Dr. A.P.J. Abdul Kalam unveiled India 2020: A Vision for the New Millennium (Kalam and Rajan, 2002), which calls for developing an India that is free from poverty, strong in trade, commerce and science and technology and that also provide healthcare and education to all.

For India to be in ranks of developed nations, it should invest in its people, create infrastructure and grow the economy. He noted that for a nation to progress, the golden triangle of R&D laboratories-academia-industry must emerge. The report conceives of India evolving into an information society and knowledge economy built on the foundations of ICTs.

New ideas, methods, skills and expertise are the intangible assets of a knowledge economy. The growth of knowledge economy hinges on the nature and extent of human capital formation of a nation. The most facile method to form human capital and hence knowledge accumulation is to invest in education and training of human resource. Since independence, in India a greater attention has been given to human capital formation by investing in education, research and training.

# 10

# Summing Up

For the last over 200 years, Neo-Classical economists had recognized only two factors of production: labour and capital. This is now changing. Information and Knowledge are replacing capital and energy as primary wealth assets, just as the latter replaced land and labour 200 years ago. Technological developments in the 20th century have transformed the majority of wealth creating work form physically- based to "knowledge-based". In both developing and industrial countries, there is an increasing awareness of the importance of knowledge, and modern developed societies have become information based knowledge societies. New ideas, methods, skills, R&D, and learning-by-doing are intangible assets of a knowledge economy. Investment in knowledge building through education, and research and development (R&D) has further accentuated knowledge accumulation at a still accelerated rate. The development and integration of new knowledge leads to successful innovation. We are now an information society in a knowledge economy where knowledge management is essential.

Since early 1980s, and more specifically since early 1990s, the whole world is integrating as a global market economy, characterized by stimulated and unrestrained trade, capital and technology flows across geographical frontiers of countries. Globalization has induced the economies to exchange based on comparative advantage. Globalization not only leads to specialization and spatial concentration of commodity production, it also requires large sophisticated service industries. Globalization has made knowledge of world much more easily available than ever before. This changing economic scenario has far-reaching implications for employment, productivity, and efficiency.

The process of globalization has led to the spread of new technologies across the world, particularly in the form of ICT, which has become the new buzzword for economic development both at global and local levels. Globalization, multilateralism and economic integration have led the factor and labour markets to converge. Information and Communication Technologies (ICT) has quickened the pace of exchange and transactions. Thus, it has proved to be the vehicle of global integration. On the other hand, the process of globalization itself registered unhindered development of ICT, hence a symbiotic relationship between globalization and spread of knowledge economy through the use of ICT has come to be strongly entranced in the global economic dynamic. Thus, ICT has set in an unprecedented growth of economic trends and commodity/factor price convergence across the global markets.

The 1990s have seen progressive integration of Indian economy into global economy. In the early 1990s, liberalization of investment, and foreign exchange regime stimulated industrial and service growth. With increase in the use of ICT in knowledge intensive sectors of the economy, India has a tremendous growth impetus to emerge as the fourth largest economy in the world. India's telecommunication sector has registered rapid growth in recent years, spurred by economic reforms. India has largely developed around three functional pillars of the knowledge economy i.e., education, innovation, and ICTs (Dahlman and Utz, 2005).

Indian ICT sector consist of information technology (IT) and telecommunication. According to National Association of Software and Service Commission (NASSCOM), Indian IT industry is broadly categorized into three segments: IT services and software; IT-enabled services (ITES)-business processing outsourcing (BPO); and hardware segment. Telecommunication is another pillar of India's ICT sector. The Indian telecom market may be broadly categorized into telephony and internet services. Telephony services can be further divided into wire line (fixed-line services) and wireless (cellular and WLL).

Indian IT industry is of recent origin, but it has attained an important place in the economy. The industry's contribution to national economic output has escalated nearly by six times, from less than 1 percent in 1991-92 to 6.46 percent in 2006-07. India enjoys enormous comparative advantage in the export of IT services. The rapid growth of the industry has been attributed to availability of human capital, entrepreneurship, social networking, favourable policy environment and entry of multinationals. India's strong education framework ensured ample supply of technical and non-technical talent, while the establishment of Software Technology Parks of India (STPI) and later SEZs provided an enabling environment for the industry to flourish.

The most important physical infrastructure needs of IT are power and telecommunications, but there is significant dependence on urban infrastructure including road transport as well. The personnel need to move speedily from their homes to workplace and they require good urban infrastructure for providing them with quality of life. Proximity to airports and good international air connections are also critical to enable them to travel frequently to foreign destinations. With all these facilities the industry requires office space. Big players need developed land to construct their own buildings but others need rented accommodation provided by developers in which they can move quickly. The end of public sector monopoly in telecommunications—resulting in large capacity creation by the private sector and significant decline in tariffs coupled with improvement in service quality—has been one of the factors contributing to the growth of IT industry. Large capacity expansion by a number of players has established a strong framework of planned redundancies, thus ensuring continuous network connectivity.

The Indian IT industry has steadily moved up the value

chain and is now fully capable of envisioning and developing a whole range of value added services that can be delivered through computers, mobile phones or other devices. The industry can create and sustain a service delivery platform for such services thereby extending the global reach of the entire service industry. The advent of cloud computing technology has thrown up another wide range of possibilities for India. Indian companies have already begun creating new products and services incorporating their own IPs and patents. The stage is now set for a quantum jump in this trend. Another positive aspect is that there is a growing trend of entrepreneurship and product/service innovation. Information technology (IT) and telecommunications are the twin sectors that best epitomize what modern, resurgent and young India is capable of accomplishing. Achievements in the recent past in the IT and telecom sectors have been spectacular and unprecedented in India's history. Yet, it is the future that holds even more breathtaking possibilities. A fascinating opportunity has now arisen that rests on the foundations of what has been achieved so far in IT and telecom.

# References

# References


Aggarwal, A. (2009), *Impact Assessment of Sunset Clause, SEZ Policy and the Global Recession on STP Units*, New Delhi: STPI.

Ahluwalia, I. J. (1991), *Productivity and Growth in Indian Manufacturing*, New Delhi: Oxford University Press.

Archibald, G.C. and G. Rosenbluth (1975), "The New Theory of Consumer Demand and Monopolistic Competition", *Quarterly Journal of Economics*, Vol. 89, pp. 569-590.

Arora, A., V.S. Arunachalan, J. Asundi and R. Fernandes (2002), "The Indian Software Services Industry", *Research Policy*, Vol. 30, No. 3, pp. 1267-1287.

Autor, D.H., F. Katz and A. Kruger (1998), "Computing Inequality: Have Computers Changed the Labour Market", *Quarterly journal of Economics,* Vol. 113, No. 4, pp. 1169-1213.

Autor, D.H., F. Levy and R.J. Murane (2003), "Skills Content of Recent Technological Charge: An Empirical Exploration", The *Quarterly Journal of Economics,* Vol. CXVIII, No. 4, p. 1280.

Baijlal, P. and R. Jain (2007), "Rural Telecom", *India Infrastructure Report 2007*, New Delhi: Oxford University Press.

Baily, M.N. and R.Z. Lawrence (2001), "Do We Have a New Economy?", *American Economic Review,* Vol. 91, No. 2, pp. 308-312.

Bains, J.S. (1979), *Industrial Organization*, New York: Harvard Economic Series.

Balakrishan, P. (2006), "Benign Neglect or Strategic Intent?: Contested Lineage of Indian Software Industry", *Economic and Political Weekly*, Vol. 41, No. 36.

Balakrishanan, P. and M. Suresh Babu (2003), "Growth and Distribution in Indian Industry in the 1990s", *Economic and Political Weekly*, Vol. 38, No. 38, pp. 3997-4005.

Barro, R. (2001), "Human Capital: Growth, History and Policy – A Session to Honour Stanley Engermen, Human Capital and Growth", *American Economic Review: Papers and Proceedings,* Vol. 91, No. 2, p. 13.

Bauara, A., C.H. Kriebel and T. Mukhopadhyay (1995), "Information Technology and Business Value: An Analytic and Empirical Investigation". (Original not seen. Cited by Dedrick, J., V. Gurbaxani and K.L. Kraemer, 2003, "Information Technology and Economic Performance: A Critical Review of Empirical Evidence", *ACM Computing Surveys*, Vol. 35, No. 1, pp. 1-28).

Bernard, A. and B. Jensen (1999), "Exceptional Exporters performance: Cause, Effect or Both?", *Journal of International Economics*, Vol. 47, No. 1, pp. 1-25.

Berndt, E.R., C.J. Morrison and L.S. Rosenblum (1992), "High-Tech Capital, Economic Performance and Labour Composition in US Manufacturing Industries: An Empirical Analysis", MIT Working Paper 3414EFA.

Bezborah, P. (2006), "Knowledge in Globalizing Economy", *Yojana*, February, p. 12.

Bhardwaj, R. (2006), "Knowledge in a Globalizing Economy", *Yojana,* February, p. 5.

Bhatt, M.S. and A. Illiyan (2009), *Information Technology (IT) in the Indian Economy: Policies, Prospects and Challenges*, New Delhi: New Century Publication.

Bishop, R.L. ((1952), "Elasticities, Cross Elasticities and Market Relationships", *American Economic Review*, Vol. 42, pp. 779-803.

Bosworth, B.P. and J.E. Triplett (2000), "What's New About the New Economy? IT, Economic Growth and Productivity, Working Paper, Brookings Institution, Washington , D.C.

Bressnahan, T.F., E. Brynjolfsson, and L. M. Hitt (2002), "Information Technology, Workplace Organization, and Demand for Skilled Labour: Firm Level Evidence", *The Quarterly Journal of Economics,* Vol. CXVII, No. 1 p. 34.

Brynjolfsson, E. (1996), "The Contribution of Information Technology to Consumer Welfare" *Information Systems Research,* Vol. 7, No. 3, pp. 281-300.

Brynjolfsson, E. and L.M. Hitt (1995), "Information Technology as a Factor of Production: The Role of Differences among Firms", *Economics* of *Innovation* and *New Technology*, Vol. 3. No. 4, pp. 183-200.

Brynjolfsson, E. and L.M. Hitt (1996), "Paradox Lost? Firm Level Evidence on the Returns to Information System Spending", *Management Science*, Vol. 42, No. 4, pp. 541-558.

Brynjolfsson, E. and L.M. Hitt (1998), "Beyond the Productivity Paradox: Computers are Catalyst for Bigger Change", *Communication of ACM,* Vol. 41. No. 8, pp. 49-55.

Brynjolfsson, E. and L.M. Hitt (2000), "Beyond Computation: Information Technology, Organizational Transformation and Business Performance", *Journal of Economic Perspective*, Vol. 52, No. 4, pp 23-48.

Brynjolfsson, E. and L.M. Hitt and S. Yang (2000), Beyond Computation: Information Technology, Organizational Transformation and Business Performance", *Journal of Economic Perspective*, Vol. 14, No. 4, pp. 23-48.

Buckley, P. J., J.H. Dunning and R.D. Pearce (1978), "The Influence of Firm Size, Industry, Nationality and Degree of Multi-nationality on the Growth and Profitability of the World's Largest Firms 1962-1972", *Weltwirtschaftliches Archives,* Vol. 114, No. 2, pp. 243-257, retrieved from www.archive.org.

Burange, L.G. (1999), "Industrial Growth and Structure: Manufacturing Sector in Maharashtra", *Economic and Political Weekly*, Vol. 34, No. 9, pp. M 39-M 48.

Caselli, F. and W.J. Coleman II (2001), "Cross-Country Technology Diffusion: The Case of Computers", *American Economic Review*, Vol. 91, pp. 328-335.

Caves, R.E. (1992), *Industrial Efficiency in Six Nations*, Cambridge: MIT Press, Massachusetts.

Centre for Monitoring Indian Economy (CMIE), (2007), *Industry: Financial Aggregates and Ratios*, Mumbai: CMIE, Economic Intelligence Service.

Centre for Monitoring Indian Economy (CMIE), (2007), *Industry: Market Size and Shares,* Mumbai: CMIE, Economic Intelligence Service.

Centre for Monitoring Indian Economy (CMIE), (2008), Infrastructure, Mumbai: Economic Intelligence Service, p. 67.

Chadha, V. (2006), "Employment of Scientific and Technical (S&T) Workforce in an Emerging Knowledge Economy Like India: The Case of Some Selected Science Intensive Sectors", *Labour and Development*, Vol. 11-12, No. 1 and 2, pp. 133-149.

Chadha, V. (2008), "Stock of Technical Manpower in India: Prospects for Employment in the Emerging Labour Market", *The Indian Journal of Labour Economics*, Vol. 51, No. 4, pp. 831-842.

Chakraborty, C. and C. Jayachandran (2001), "Software Sector: Trends and Constraints", *Economic and Political Weekly,* Vol. XXXVI, No. 34, p. 3257.

Chandershekher, S. and K.P. Basvarajappa (2001), "Technological Innovation and Economic Development: Choice and Challenges for India" *Economic and Political Weekly*, Vol. 36, No. 34, pp. 3238-3245.

Chaudhuri, D.D. (1995), "Technological Capability in Indian Electronics Industry Under Economic Liberalization", *Economic and Political Weekly*, Vol. 30, No. 7 and 8, pp. M 13-M 16.

Chowdhary, T.H. (2002), "Telecom: Policy Response to Change", *Economic and Political Weekly,* Vol. 37, No. 32, pp. 3317-3318.

Collins, N.R. and L.E. Peterson (1970), "Concentration and Price-Cost Margins in Manufacturing", (Original not seen.

Cited by Hay, D.A. and D.J. Morris, 1975, *Industrial Economics: Theory and Evidence*, London: Oxford University Press.

Comanor, W.S. and T.A. Wilson (1967), "Advertising, Market Structure and Performance", *Review of Economics and Statistics*, Vol. 49, pp. 423-440.

Council of Economic Advisor (2000), The Annual Report of the Council of Economic Advisor. (Original nor seen. Cited by Dedrick, J., V. Gurbaxani and K.L. Kraemer, 003)."Information Technology and Economic Performance: A Critical Review of Empirical Evidence", *ACM Computing Surveys*, Vol. 35, No. 1, pp. 1-28.

Cowling, K. and J. Cubbin (1971), "Price, Quality and Advertising Competition", *Economica*, Vol. 38, pp. 378-394.

Cowling, K. and M. Waterson (1976), "Price-Cost Margins and Market Structure", *Economica*, Vol. 43, pp. 267-274.

Dahlman, C. and A. Utz (2005), *India and the Knowledge Economy: Leveraging Strengths and Opportunities*, Washington: WBI Studies, World Bank.

David, D.B. and M.B. Loewy (2002), "Knowledge Dissemination, Capital Accumulation, Trade and Endogenous Growth", *Oxford Economic Papers*, Vol. 52, pp. 637-650.

David, P.A. (1990), "The Dynamo and the Computers: A Historical Perspective on the Modern Productivity Paradox", *American Economic Review: Papers and Proceedings*, Vol. 1, No. 2, pp. 355-361.

De, S. and D. Dutta (2007), "Impact of Intangible Capital on Productivity and Growth: Lessons for Indian Information Technology Software Industry", *The Economic Record*, Vol. 83, pp. S 73-S 86.

Dedrick, J., V. Gurbaxani and K.L. Kraemer (2003), "Information Technology and Economic Performance: A Critical Review of Empirical Evidence", *ACM Computing Surveys*, Vol. 35, No. 1, pp. 1-28.

Desai, A.V. (2006), *India's Telecommunication Industry: History, Analysis and Diagnosis*, New Delhi: Sage Publications.

Dewan, S. and K.L. Kreamer (2000), "Information Technology and Productivity: Preliminary Evidence from Country Level Data", *Management Science,* Vol. 46, No. 4, pp. 56-62.

Dossani, R. (2005), "Origins and Growth of Software Industry in India", Shorenstein Asia-Pacific Research centre, Working Paper.

Dutz, M.A. and C. Dahlman (2007), The Indian Context and Enabling Environment. In Dutz, A. (eds.). *Unleashing India's Innovation: towards Sustainable and Inclusive Growth*, Washington: The World Bank.

Dutta, S., B. Lanvin and F. Paua (2003), *The Global Information Report 2002-03: Readiness for Networked World*, New York and Oxford: Oxford University Press.

Eaton, J. and S. Kortum (1996), "Trade in Ideas, Patenting and Productivity in the OECD", *Journal of International Economics*, Vol. 40, No. 1, pp. 41-45.

Ferguson, R.W. and W.L. Wascher (2004), "Distinguished Lecture on Economics in Government: Lessons from Past Productivity Booms", *Journal of Economic Perspectives,* Vol. 18, No. 2, p. 19.

FICCI and CSIR (2006), "Think Innovation, Think India", *Yojana*, February, pp. 27-35.

Freund, C. and D. Weinhold (2002), "The Internet and International Trade in Services", *AEA Papers and Proceedings,* Vol. 92, No. 2, p. 237.

Gangadhar, V. and G.N. Reddy (2005), "Strategies for Business Process Outsourcing in India", *Economic and Political Weekly*' Vol. 54, No. 6, pp. 857-865.

Garbade, K.D. and W.L. Silber (1978), "Technology, Communication and the Performance of Financial Markets: 1840-1975", *Journal of Finance*, Vol. 33, No. 3, pp. 819-832.

Gayathri, V. and M. Vijaybaskar (2003), "Information and Communication Technology and Indian Development: A Report on International Seminar", *The Indian Journal of Labour Economics*, Vol. 46, No. 1.

Gilchrist, S., V. Gurbaxani and R. Town (2001), "PCs and the Productivity Revolution", Working Paper, University of California: Centre for Research on Information Technology and Organization, Irvine, CA.

Goldar, B.N. (2002), "Total Factor Productivity Growth in Indian Manufacturing in 1980s", *Economic and Political Weekly*, Vol. 37, No. 49, pp. 4966-4968.

Goldar, B.N. (2004), "Indian Manufacturing: Productivity Trends in Pre- and Post-Reform Periods", *Economic and Political Weekly*, Vol. 39, No. 46 and 47, p. 5035.

Gordon, R. (2000), Does the "New Economy" Measure Up to the Great Inventions of the Past?", *Journal of Economic Perspective*, Vol. 14, No. 4, pp. 49-76.

Government of India (1986), *Policy on Computer Software Exports, Software Development and Training*, New Delhi: Department of Electronics.

——Government of India (2002a), *India Vision 2020*, retrieved fromhttp://planningcommission.nic.in/reports/genrep/pl_vsn2020pdf.

——(2002b), *Tenth Five Year Plan: Information Technology Sector*, New Delhi: Ministry of Commerce and Information Technology, Department of Information and Technology.

——(2004), *Telecommunication Statistics*, New Delhi: Ministry of Communication and Information Technology, Department of Telecommunication.

——(2005a), *Advantage India,* New Delhi: Ministry of Commerce and Information Technology, Department of Information and Technology.

——(2006a), *Electronics Information and Planning*, Vol. 34, No. 1-2.

——(2007a), *Electronics Information and Planning*, Vol. 34,

No. 5-6.

——(2007b), *Electronics Information and Planning*, Vol. 34, No. 9-10.

——(2007c), *Electronics Information and Planning*, Vol. 34, No. 3-4.

——*Annual Report*, New Delhi: Ministry of Communication and Information Technology, Department of Information Technology.

——*Annual Report,* New Delhi: Ministry of Commerce and Information Technology, Department of Telecommunication.

Goyal, N. K. (2008), *Indian Telecom*, retrieved from http://www.nkgoyals.com/telecom.aspx.

Grossman, G. M. and E. Helpman (1990), "Comparative Advantage and Long Run Growth", *American Economic Review*, Vol. 80, No. 4, pp. 796-825.

Grossman, G. M and E. Helpman, (1991), "Quality Ladders in the Theory of Growth," *Review of Economic Studies*, Vol. 58, No. 1, pp. 43-61.

Gupta, A., P. Mishra and P. Chakraborty (2007), *India Overall Trends in IT Spending by Industry Verticals, 2007-2011: Forecast and Analyst*, Gurgaon: IDC.

Gupta, S.C. (2010), *Fundamentals of Statistics*, Mumbai: Himalaya Publishing House, p. 670.

Hajara, S. (1965), "Cost-Price Relationship in India n Manufacturing Industries", The Economic weekly, October 23, pp. 1629-1636.

Harley, C.K. (2000), "A Review of O' Rourke and Williamson's Globalization and History: The Evolution of a Nineteenth Century, Atlantic Economy", *Journal of Economic Literature*, Vol. XXXVIII, p. 933.

Harris, S.E. and J.L. Katz (1991), "Organizational Performance and Information Technology Investment in the Insurance Industry". (Original not seen. Cited by Dedrick, J., V. Gurbaxani and K.L. Kraemer, 2003, "Information Technology and Economic Performance: A

Critical Review of Empirical Evidence", *ACM Computing Surveys*, Vol. 35, No. 1, pp. 1-28.

Hay, D.A. and D.J. Morris (1979), *Industrial Economics: Theory and Evidence*, London: Oxford University Press.

Heeks, R. (1995), Import Liberalization and Development of Indian Computer Industry", *Economic and Political Weekly,* Vol. 30, No. 34, pp. M 82-M 93.

Heeks, R. (1996), *India's Software Industry: State Policy, Liberalization and Industrial Development,* New Delhi: Sage Publications.

Hoeven, R.V.D. (2010), "Labour Markets Trends, Financial Globalization and the Current Crisis in Developing Countries", DESA Working Paper No. 99 ST/ESA/2010/DWP/99 retrieved from http://www.un.org/esa/desa/papers/2010/WP99.2010.pdf., accessed on 21st October, 2010.

Hutchison, F. and P.V. Ilavarasan (2008), "The IT/ITES Sector and Economic Policy at the Sub-National Level in India", *Economic and Political Weekly,* Vol. XLIII, No. 15, pp. 64-70.

International Data Corporation (2009), *India IT/ITeS industry expected to cross US$ 110 Billion by 2013, 5–year CAGR at 13.9 percent: Domestic IT market to grow faster than exports*, Press release June 29, retrieved from www.idc.org.

ILO (2001), *World Employment Report-2001*, Geneva, ILO.

Iyengar, J. (2005), "3G: Spectrum Issues", *Chartered Financial Analyst,* August, pp. 45-46.

Jayram, A. and B. Kalesh (2007), "Slow Nation", *Business World,* September 3, pp. 25-31.

Jorgenson, D.W. (2001),"Information Technology and the U.S. Economy", *American Economic Review,* Vol. 91, No. 1. pp. 1-32.

Jorgenson, D.W. and K.J. Stiroh (1999), "Productivity Growth: Recent Recovery and Longer Term Trends", *The*

*American Economic Review*, Vol. 89, No. 2, pp. 109-115.

Joseph, K J. (1997), *Industry under Economic Liberalization: The Case of Indian Electronics*, New Delhi: Sage Publications.

Joseph, K.J. (2006), *Information Technology, Innovation System and Trade Regime in Developing Countries*, New York: Palgrave McMillan.

Joseph, K.J. (2009), "IT Industry in India: Leading Software and Lagging Hardware". In Bhatt, M. S. and A. Illiyan (eds.). *Information Technology in the Indian Economy*, New Delhi: New Century Publications

Joseph, K.J. and K.N. Harilal (2001), "Structure and Growth of India's IT Exports: Implications of an Export Oriented Growth Strategy", *Economic and Political Weekly,* Vol. 36, No. 34, p. 3265.

Kalam, Abdul A.J.P. (2006), "Innovations is the Key", *Yojana,* Feb, pp. 2.

Kalam, A.P.J. Abdul and Y.S. Rajan (2002), *India 2020: A Vision for the New Millennium*, New Delhi: Penguin Book India.

Katrak, H. (1980), "Cost-Price Relationship in Indian Manufacturing Industries", *The Journal of Development Studies*, Vol. 17, pp. 62-79.

Keller, W. (2004), "International Technology Diffusion" *Journal of Economic Literature,* Vol. XLII, p. 768.

Khalilzadeh-Shirazi, J. (1974), "Cost-Price Relationship in Indian Manufacturing Industries", *The Economic Weekly*, Oct. 23, pp. 1629-1636.

Khan, M.Y. and P.K. Jain (2000), *Management Accountant*, New Delhi: Tata McGraw-Hill Publishing Co. Ltd.

Krueger, A.B. (1991), "How Computers have Changed the Wage Structure: Evidence from Micro-data-1984-89", Working Paper No. 3858, National Bureau of Economic research, Cambridge.

Krugman, P. (1987), "The Narrow Moving Band, the Dutch Disease and the Consequences of Mrs. Thatcher Notes on

Trade in the Presence of Scale Economies", *Journal of Development Economics*", Vol. 27, No. 1, pp. 41-55.

Krugman, P. (1994), "Past and Prospective Causes of High Unemployment", *Economic Review*, No. 4.

Kumar, N. (2001), "Indian Software Industry Development – National and International Perspective", *Economic & Political weekly,* Vol. 36, No. 45, p. 4284.

Kumar, N. and K.J. Joseph (2005), "Export of Software and Business Process Outsourcing from Developing Countries: Lessons from the Indian Experience", *Asia Pacific Trade and Investment Review*, Vol. 1, No. 1, pp. 91-110.

Kumar, T.S. and M.S. Kumar (2008), Economic Reforms and Total Factor Productivity Growth in Southern States of India", *Indian Journal of Economics*, Vol. LXXXIX, No. 4, pp. 537-555.

Lal, K. (1996), "Information Technology, International Orientation and Performance: A Case Study of Electrical and Electronic Goods Manufacturing Firms in India", *Information Economics and Policy (North-Holland*), Vol. 8, No. 3, pp. 269-280.

Lal, K. (1998), "The Adoption of Information Technology and its Consequences: A Case Study of Indian Manufacturing", *Science Technology and Development*, Vol. 16, No. 1, pp. 81-100.

Lal, K. (2001), The Determinants of the Adoption of Information Technology: A Case study of the Indian Garment Industry. In Pojola, M. (eds.). *Information Technology, Productivity and Economic Growth: International Evidence and Implications for Economic Development,* Cambridge: Oxford University Press, U.K., pp. 149-174.

Lal, K. (2002), "E-Business and Export Behaviour", WIDER Discussion Paper, 2002/68.

Loveman, G. W. (1994), An Assessment of the Productivity Impact of Information Technology. In Allen, T. J. and M. S. Scott Morton (eds.). *Information Technology and the*

*Corporation of the 1990s: Research* Cambridge: Oxford University Press, U.K., pp. 84-110.

Lowe, P. (1995), "Labour productivity Growth and Relative Wages: 1978-1994", Reserve Bank of Australia Conference on Productivity and Growth.

Mancke, R.B. 991974), "Causes of Inter-firm Profitability Differences", *Quarterly Journal of Economics,* Vol. 88, pp. 181-193.

Mani S. (2007), "Revolution in India's Telecom Industry", *Economic and Political Weekly*, Vol. XLII, No. 7, p. 579.

Mani, S. (2008), "Growth of India's Telecom Industry", *Economic and Political Weekly*, Vol. XLIII, No. 3, pp. 37-46.

Mann, M. (1966), "Seller Concentration, Barriers to Entry and Rates of Return in 30 Industries", *Review of Economics and Statistics,* Vol. 48, pp. 296-307.

Mariappan, V. and K. Chidambaram (2003), Public Sector Textile Mills: Productivity Performance", *Economic and Political Weekly*, Vol. 38, No. 16, pp. 1551-1554.

Mathur, S.K. (2006), "Indian Information Technology Industry: Past Present and Future and a Tool for National Development", *Journal of Theoretical and Applied Information Technology*, Vol. 2, No. 2.

McKinsey Quarterly (2004), *Off-shoring Goes on the Offencing: Cost Cutting is only the First Benefit*, Number 2.

Mc Millin, D. (2006), "Outsourcing Identities: Call Centres and Cultural Transformation in India" , *Economic and Political Weekly,* Vol. 41, Jan. 21-26, p. 237.

Mohan, D. (2009), "Log-in Failure?", *Business India*, 8th March, pp. 55-56.

Moodley, A. (2002), "Competing in the Digital Economy? The Dynamics and Impact of B2B E-Commerce on the South-African Manufacturing Sector, WIDER Discussion Paper, 2002/79, retrieved from www.wider.unu.edu/publications /working-papers/discussion-papers/en_GB/discussion-papers/.

Muhammad, A. (2006), "Globalization ad Economic Transformation in a Peripheral Economy", *Economic and Political Weekly,* Vol. XLI, No. 15, p. 1459.

Murphy, K.M., W.C. Riddell and P.M. Romer (1998), "Wages, Skill and Technology in the U.S and Canada", Working Paper No. 6638, Cambridge: National Bureau of Economic Research, Cambridge.

Murphy, R.V.R. and D.K. Bahera (2008),"Employment Potential in India's Service Sector in the Post-Liberalization Period in India", Indian Journal of Economics, Vol. 89, no. 4, pp. 599-609.

Naik, S.D. (2001), "IT Hardware: A Costly Neglect", *Economic and Political Weekly,* Vol. 36, No. 27, pp. 2481.

Narayanan, K.S. (2007), "Seeking Internal Gurus", *Business India,* No. 769, pp. 112

NASSCOM (2005a), "NASSCOM Announces Ranking of Third-Party ITES", *Yojana,* Vol. 49, p. 18.

NASSCOM (various issues), *Strategic Review: IT Industry in India,* New Delhi: NASSCOM.

NASSCOM- A. T. Kearney (2008), *India: Location Roadmap for IT-BPO Growth,* New Delhi: NASSCOM.

NASSCOM- McKinsey (2005), Extending India's Leadership of the Global IT and BPO Industries, New Delhi: NASSCOM.

Navaretti, G.B. and D.G. Tarr (2000), "International Knowledge Flows and Economic Performance: A review of Evidence", *The World Bank Economic Review,* Vol. 14, No.1, p. 04.

Oliner S.D. and D.E. Sichel (2000), "The Resurgence of Growth in the late 1990: Is Information Technology the Story?" *Journal of Economic Perspective*, Vol. 14, No. 4, p. 18.

Oliner, S.D. and D.E. Sichel (2002), "Information Technology and Productivity: Where are We and Where are We Going?", *Economic Review*, Vol. 3, No. 3, pp. 15-41.

OECD (2005), http://www.oecd.org/49/44/35930616.pdf, accessed in January 2012.

Pandey, I.M. (2010), *Financial Management*, New Delhi: Vikas Publishing House Pvt. Ltd, pp, 581-611.

Pohjola, M. (2001), Information Technology and Economic Growth: A Cross Country Analysis. In Pohjola, M. (eds.). *Information Technology and Economic Development*, Cambridge: Oxford University Press, U.K., pp. 242 - 256.

Prasuna, D.G. (2003), "BPO: Advantage India", *Chartered Financial Analyst*, January, pp. 19-26.

Rai, S. (2005), "Companies Rebel over Bangalore Sad State", *International Herald Tribune*, September, 13.

Raipuria, K. (2002), "What Size the New Economy: A Conduit Approach", *Economic and Political Weekly*, Vol. 37, March 16, pp. 1062-1067.

Rajendran, M. (2008), "Missed Opportunity", *Business World*, Vol. 28, No. 28, pp. 46-47.

Ramachandriah, C. (2003), "Information Technology and Social Development", Economic *and Political Weekly*, Vol. 38, No. 12, pp. 1192-1197.

Ramani, S. (2000), "IT Enabled Services: Growing Forms of Telework", *Economic and Political Weekly*, Vol. 35, No. 26, pp. 2305-2308.

Ramaswamy, K.V. (1994), "Small-Scale Manufacturing Industries: Some Aspects of Size, Growth and Structure", Economic *and Political Weekly*, Vol. 29, No. 9.

Ramesh, B.P. (2004), "Cyber Coolies' in BPO: Insecurities and Vulnerabilities of Non – Standard Work", *Economic and Political Weekly*, Vol. XXXIX, January 31, p. 492.

Ramesh, B.P. (2009), Economic Upgrading in India's IT/IT-Enabled and BPO Sector. In Bhatt, M. S. and A. Illiyan (eds). *Information Technology in the Indian Economy*, New Delhi: New Century Publications.

Ramesh, T. (2005), "Outsourcing-A New Wave of Business", *The Management Accountant*, Vol. 40, No. 9, pp. 691-696.

Rao, C. S. (2005), "Information and Communication", *Margin*,

Vol. 38, No. 1, pp. 59-80.

Rao, D.N. (2009), Tardy Growth of IT Hardware Industry in India. In Bhatt, M.S. and A. Illiyan (eds.). *Information Technology in Indian Economy*, New Delhi: New Century Publications.

Rao, N.J. and S. Kumar (2005), "BPO Bash-Is India Loosing Out", *Chartered Financial Analyst*, June, pp. 58-61.

Ray, S. (2004), "MNE Strategic Alliances and Efficiency of Firms: Emerging Trends", *Economic and Political Weekly,* Vol. 39, No. 5, pp. 434-440.

Ribeiro, J. (2005), "IT Industry Protest Bangalore Infrastructure", *Infoworld,* September 12, retrieved from www.infoworld.com.

Rivera-Batiz, L. and P.M. Romer (1991a), "Economic Integration and Endogenous Growth", *Quarterly Journal of Economics*, Vol. 106, No. 2, pp. 531-55.

Roach, S.S. (1987), "America's Technology Dilemma: A Profile of the Information Economies", Morgan Stanly Special Economic Study, April.

Rohini, S. (2004), "Steel Industry: A Performance Analysis", *Economic and Political Weekly,* Vol. 39, No. 16, pp. 1613-1620.

Roller, L. H. and L. Waverman (2001), "Telecommunication Infrastructure and Economic Development: A Simultaneous Approach", *American Economic Review,* Vol. 91, No. 4, pp 909-923.

Rowthorn, R. (1971), *International Big Business 1957-1967: A Study of Comparative Growth,* Cambridge: Cambridge University Press.

Roy, A. (2003), "Indian IT Industry: The Second Wave", *Chartered Financial Analyst*, September, p. 25.

Saha, M. (2005), "Human Resource Outsourcing: A Strategy for Gaining Competitive Advantage", *Economic and Political Weekly*, Vol. 54, No. 6, pp. 866-874.

Sandesara, J.C. (1993), "Modern Small Industry, 1972 and 1987-88: Aspects of Growth and Structural Change",

*Economic and Political Weekly,* Vol. 33, No. 6, pp. 223-229.

Schware, R. (1987), "Software Industry in the Third World: Policy Guidelines, Institutional Options and Constraints", *World Development*, Vol. 15, No. 10/11, pp. 1249-67.

Scott, A.J. (1996), "Regional Motors of the Global Economy", *Fortunes*, Vol. 28, No. 5, pp. 391-411.

Seigel, D. (1997), "The Impact of Computers on Manufacturing Productivity Growth: A Multiple-Indicators, Multiple-Cause Approach", *The Review of Economic and Statistics*, Vol. 79, No. 1, pp. 68-78.

Sen, P. (1995), "Indian Software Exports: An Assessment", *Economic and Political Weekly,* Vol. 30, No. 7 and 8, pp. 2053-2058.

Sengupta, D.N. (1993), "Indian Manufacturing Industry: Growth Episode of Eighties", *Economic and Political Weekly,* Vol. 38, No. 22, pp. M 54-M 62.

Sharma, S. (2011), "100 Million Jobs on the Envil", *The Tribune*, June 10.

Sharma, S. and L. Singla (2009), "Telecom Equipment Industry: Challenges and Prospects", *Economic and Political Weekly,* Vol. XLIV, No. 1, pp. 16-17.

Sibal, K. (2008), "India Lags China in Research and Development Spending", *The Financial Express,* 13th March.

Siddarthan, N.S. (1988), "Technology, Modernization and Growth: A Case Study of Indian Corporate Sector 1975-83", *Economic and Political Weekly*, Vol. 23, No. 31, pp. 1587-1590.

Siddarthan, N.S. (1999), "European and Japanese Affiliates in India: Difference in Conduct and Performance", *Economic and Political Weekly*, Vol. 34, No. 22, pp.M61-M65.

Siddarthan, N.S. and A. Dasgupta (1983), "Entry Barrier, Exports and Inter Industry Differences in Profitability", *Developing Economies*, Vol. 21, pp. 14-23.

Siddarthan, N.S. and K. Lal (1988), "The Recent growth of the

Largest US Multinational", *Oxford Bulletin of Economic and Statistics*, Vol. 44, pp. 1-13.

Siddarthan, N.S. and K. Lal (2003), "Liberalization and Growth of Firms in India", *Economic and Political Weekly,* Vol. 38, No. 20, pp. 1983-1987.

Siddarthan, N.S. and K. Lal (2004), "Liberalization, MNE and Productivity of Indian Enterprises", *Economic and Political Weekly*, Vol. 39, No. 5, pp. 448-452.

Sidhu, H. (2010), "Productivity Led Wage Disparity in the Indian Industry", *Indian Journal of Industrial Relations*, Vol. 45, No. 3, pp. 350-366.

Sikidar, S. and R.K. Raul (2005), "Business Process Outsourcing: An Analytical Study", *The Management Accountant*, Vol. 40, No. 9, pp. 686-690.

Singh, J. (2008), "Domestic IT Market Outperform Exports", The Tribune, 9th September, p. 17.

Srivastava, K.B.L. (2001), "Intellectual Capital Development and Management of Knowledge is Knowledge Based Economy", *Indian Journal of Industrial Relations,* Vol. 36, No. 3, p. 356.

Stigler, G.J. (1963), *Capital and Rates of Return in Manufacturing Industries*, Princeton: Princeton University Press.

Stiglitz, G.J. (2003), "Globalization, Technology and Asian Development", *Asian Development Review,* Vol. 20, No. 2, p. 03.

Stiroh, K.J. (2002), "Information Technology and the U.S. Productivity Revival: What do the Industry Data Say", *American Economic Review*, Vol. 92, No. 5, p. 1574.

Stokey, N. (1991), "Human Capital, Product Quality and Growth", *Quarterly Journal of Economics*, Vol. 106, No. 2, pp. 587-616.

Stonebraker, R. J. (1976), "Corporate Profits and the Risk of Entry", *Review of Economics and Statistics*, Vol. 58, pp. 33-39.

Strassman, P.A. (1990), *The Business Value of Computers: An*

*Executive's Guide*, Information Economic Press, New Canaan, CT.

Tagnas, R.A.L. and V. K. Kaul (2006), "Innovation System in India's IT Industry: An Empirical Investigation", *Economic and Political Weekly*, Vol. XLI, No. 39, p. 4178.

Tandon, A. (2010), "By 2020, 42 Million Students Will Need Rs. 1.6 Lakh Crore Loans", *The Tribune*, 1st July.

Telephone Regulatory Authority of India (TRAI), (2005b), *Recommendations on the Growth of Telecom Services in Rural India*, New Delhi: TRAI.

Telephone Regulatory Authority of India (TRAI), (2008), *Annual Report 2006-07,* Ministry of Communication and Information Technology, retrieved from www.trai.gov.in.trai_annualreport.asp.

The Economist (2004a), "A Survey of Outsourcing", Vol. 373, No. 8401, pp. 3-12.

The Economist (2004b), "The place to be-A Survey of Outsourcing", November 13, p. 8.

The Economist (2005a), "Outsourcing and IT in India, Vol. 375, No. 8423, pp. 65-68.

The Economist (2005b), "The Real Digital Divide", March 10, retrieved from http://www.economist.com/printedition/displaystory.cfm?story_D=3742817 on 12th August, 2005.

Thomas, J.J. (2005), New Technologies for India's Development. In Parikh, K.S. and R. Radhakrishanan (eds). *India Development Report 2004-05,* New Delhi: Oxford University Press, pp. 126-134.

Timmer, M.P. and B.V. Ark (2005), "Does Information and Communication Drive EU-US Productivity Growth Differentials?" *Oxford Economic Papers,* Vol. 57, No. 4, p. 694.

Tybout, R. James (1992), "Linking Trade and Productivity: New Research Directions", *The World Bank Economic Review,* Vol. 6, No. 2, pp. 189-211.

Udo, G.J. and F.M. Edoho (2000), "Information Technology Transfer to African Nations: An Econometric Development Mandate", *Journal of Technology Transfer*, Vol. 25, pp. 329-342.

UNDP (2001), *Human Development Report 2001: Making New Technologies Work for Human Development*, New York: Oxford University Press, p. 38.

Unni, J., N. Lalitha and U. Rani (2001), "Economic Reforms and Productivity Trends in Indian Manufacturing", *Economic and Political Weekly*, Vol. 36, No. 41, pp. 1044-1050.

Van Ark, B., R.I. and R.H. Mc Guckin (2004), "ICT and Productivity in Europe and the United States: Where do the Differences Come From?", *CESifo Economic Studies*, Vol. 49, No. 3.

Veermani, C. and B. Goldar (2005), "Manufacturing Productivity in Indian States: Does Investment Climate Matter", *Economic and Political Weekly*, Vol. XL, No. 24, pp. 2413-2420.

Venkateswarlu, T. (2004), "Globalization of Business and Information Technology", *Indian Journal of Economics*, Vol. LXXXIV, No. 335, pp. 639-650.

Verma R. and E. M. Rogers (2004), "Indian Cyber Workers in US", *Economic and Political Weekly,* December 25, p. 5648.

Viren, M. (2005), "Why do Capital Intensive Companies Pay Higher Wages", Bank of Finland Research, Working Paper No. 5/2005.

Waczirag, R. (2001), "Measuring the Dynamic Gains from Trade", *The World Bank Economic Review,* Vol. 15, No. 3, p. 394.

Whelan, K. (2000), "Computers, Obsolescence and productivity: Finance and Economic Discussions services, 2000-06, Board of Governors of the Federal Reserve System

World Bank (Various Years), *Global Monitoring Report*

*(2005)*, IBRD: Washington D.C., pp. 119-141.

Yifulin, J. (2004), "Development Strategies for Inclusive Growth in Developing Asia", *Asians Development Review,* Vol. 21, No. 2, p. 11.

**Websites**

http://www.cbhidghs.nic.in
http://dqindia.ciol.com
http://education.nic.in/stats/statpub.asp
http://education.nic.in/planbudget
http://www.hdrstats.undporg
http://www.icrier.org
http://www.idc.org.
http://www.indiabudget.nic.in
http://www.ipindia.nic.in
http://www.knowledgecommission.gov.in
http://www.mospi.nic
http://www.stats.uis.unesco.org
http://www.stpi.in/index.html
http://www.un.org
http://www.unctad.org
http://www.wto.org.
http://www.who.int
http://www.uis.unesco.org
http://voicendata.ciol.com.

# Index

# Index